SpringerBriefs in Mathematics of
Planet Earth • Weather, Climate, Oceans

SpringerBriefs present concise summaries of cutting-edge research and practical applications across a wide spectrum of fields. Featuring compact volumes of 50 to 125 pages, the series covers a range of content from professional to academic. Briefs are characterized by fast, global electronic dissemination, standard publishing contracts, standardized manuscript preparation and formatting guidelines, and expedited production schedules.

Typical topics might include:

- A timely report of state-of-the art techniques
- A bridge between new research results, as published in journal articles, and a contextual literature review
- A snapshot of a hot or emerging topic
- An in-depth case study

SpringerBriefs in the Mathematics of Planet Earth showcase topics of current relevance to the Mathematics of Planet Earth. Published titles will feature both academic-inspired work and more practitioner-oriented material, with a focus on the application of recent mathematical advances from the fields of Stochastic And Deterministic Evolution Equations, Dynamical Systems, Data Assimilation, Numerical Analysis, Probability and Statistics, Computational Methods to areas such as climate prediction, numerical weather forecasting at global and regional scales, multi-scale modelling of coupled ocean-atmosphere dynamics, adaptation, mitigation and resilience to climate change, etc. This series is intended for mathematicians and other scientists with interest in the Mathematics of Planet Earth.

More information about this subseries at http://www.springer.com/series/15250

Andrew J. Majda • Samuel N. Stechmann
Shengqian Chen • H. Reed Ogrosky • Sulian Thual

Tropical Intraseasonal Variability and the Stochastic Skeleton Method

 Springer

Andrew J. Majda
Department of Mathematics, and Center for
Atmosphere Ocean Science
Courant Institute of Mathematical Sciences
New York University
New York, NY, USA

Center for Prototype Climate Modeling
NYU Abu Dhabi
Abu Dhabi, United Arab Emirates

Shengqian Chen
Nanjing University of Information Science
and Technology
Nanjing, China

Sulian Thual
Department of Atmospheric and Oceanic
Sciences
Fudan University
Shanghai, China

Samuel N. Stechmann
Department of Mathematics
University of Wisconsin-Madison
Madison, WI, USA

H. Reed Ogrosky
Department of Mathematics and Applied
Mathematics
Virginia Commonwealth University
Richmond, VA, USA

ISSN 2524-4264   ISSN 2524-4272   (electronic)
Mathematics of Planet Earth
ISSN 2509-7326   ISSN 2509-7334   (electronic)
SpringerBriefs in Mathematics of Planet Earth
Weather, Climate, Oceans
ISBN 978-3-030-22246-8   ISBN 978-3-030-22247-5   (eBook)
https://doi.org/10.1007/978-3-030-22247-5

Mathematics Subject Classification: 86xxx, 60xxx, 35xxx

This Springer imprint is published by the registered company Springer Nature Switzerland AG.
The registered company address is: Gewerbestrasse 11, 6330 Cham, Switzerland

Preface

This book should be of interest to graduate students, postdocs, and senior researchers in pure and applied mathematics, physics, engineering, and climate, atmosphere, and ocean science interested in turbulent dynamical systems as well as other complex systems.

This research is partially supported by the Office of Naval Research (ONR) through MURI N00014-16-1-2161. The first author, Andrew Majda, is Principal Investigator (PI) while Sam Stechmann was a co-PI on this grant, and they together formed a research team on the topics presented here with postdocs Shengqian "Chessy" Chen, Reed Ogrosky, and Sulian Thual. All the authors gratefully acknowledge the support, encouragement, and guidance of the ONR program managers, Reza Malek-Madani and Scott Harper, for this initiative.

New York, NY, USA   Andrew J. Majda
June 2018

Contents

1   **Introduction**   1
     References   3

2   **The Deterministic Skeleton Model and Observed Features**
     of the MJO   5
     2.1   Introduction   5
     2.2   Observed Features of the MJO   6
     2.3   The MJO Skeleton Model   9
         2.3.1   Model Description   9
         2.3.2   Energetics   13
         2.3.3   Derivation of K, R, Q, and A Variables and Evolution
                 Equations   14
     2.4   Linear Theory   18
         2.4.1   Formula for Intraseasonal Oscillation Frequency   18
         2.4.2   Linear Waves   19
     2.5   Nonlinear Simulations   21
     2.6   Nonlinear Traveling Waves: Exact Solutions   23
     2.7   Concluding Discussion   24
     References   25

3   **A Stochastic Skeleton Model for the MJO**   29
     3.1   Introduction   29
     3.2   Features of the Stochastic Skeleton Model   32
         3.2.1   Hovmoller (SpaceTime) Diagrams   32
         3.2.2   MJO Wave Trains   33
         3.2.3   Power Spectra   36
         3.2.4   Beyond the Stochastic Skeleton Model   37
     3.3   Derivation of the Stochastic Skeleton Model   39
         3.3.1   Markov Birth-Death Process   39
         3.3.2   Choice of Transition Rates   41
         3.3.3   Gillespie Algorithm   41

		3.3.4	Stochastic Single-Column Skeleton Model	42
		3.3.5	Complete Stochastic Skeleton Model	44
	3.4	Concluding Discussion		46
	References			46

4 Tropical–Extratropical Interactions and the MJO Skeleton Model 49
    4.1   Introduction 50
    4.2   Three-Wave Interactions 51
         4.2.1 The Two-Layer Equatorial Beta-Plane Equations
              with the MJO Skeleton 51
         4.2.2 The Zonal-Long-Wave-Scaled Model 53
         4.2.3 Long Time Scales for Tropical-Extratropical Interactions.... 54
         4.2.4 Asymptotic Expansions 55
         4.2.5 Eigenmodes of the Linear System 55
         4.2.6 The Reduced Asymptotic Models.............................. 56
         4.2.7 Tropical–Extratropical Interactions via Three-Wave
              Resonant Interactions... 58
    4.3   Direct Two-Wave Interactions with Climatological Mean Flow 59
    4.4   Additional Realism 62
         4.4.1 More general Walker circulation.............................. 62
    4.5   Effects of Wind Shear... 62
    4.6   Concluding Discussions 64
    References 65

**5 New Indices for Observations of Tropical Variability Based on
the Skeleton Model and a Model for the Walker Circulation** 67
    5.1   Introduction 67
    5.2   Data Analysis and Modeling of the Walker Circulation.............. 68
         5.2.1 Model Derivation 70
         5.2.2 Estimating f^θ, K, and R_m 72
         5.2.3 Results.. 74
    5.3   An MJO Index (MJOS) Based on the Skeleton Model................ 75
         5.3.1 Solutions to the Linear MJO Skeleton Model 75
         5.3.2 Data Analysis Methods... 76
         5.3.3 Results.. 77
    5.4   A Warm-Pool MJO Index (MJOS$_k$) Based on the Skeleton Model ... 79
         5.4.1 Estimating $A_s(x)$ 80
         5.4.2 Model Solutions 81
         5.4.3 Eigenvector Projection... 83
    5.5   Assessing the Validity of the Equatorial Long-Wave
        Approximation 85
         5.5.1 Meridional Winds.. 87
         5.5.2 Meridional Geostrophic Balance.............................. 87
    5.6   Conclusion 90
    References 90

**6   Refined Vertical Structure in the Stochastic Skeleton Model for
      the MJO**   93
   6.1   Introduction   93
   6.2   Skeleton Models with Refined Vertical Structure   96
      6.2.1   Representation of Convective Activities.......................   96
      6.2.2   Evolution of Convective Activities.............................   97
      6.2.3   Skeleton Model with Three Convective Activities   98
      6.2.4   Skeleton Model with Two Active Baroclinic Modes.......... 100
   6.3   Model Formulation.. 101
      6.3.1   Vertical Baroclinic Structure 102
      6.3.2   Coupling to Three Convective Activities 104
      6.3.3   Conserved Energy Principle..................................... 105
      6.3.4   Addition of a Slaved Secondary Circulation 106
      6.3.5   Active Coupling to the Second Baroclinic Mode 107
      6.3.6   A Suite of Extended Skeleton Models 108
   6.4   Discussion 110
   References 111

7   Current and Future Research Perspectives 113
   7.1   Future Directions for Stochastic Skeleton Models 113
      7.1.1   Data Assimilation and Real-Time Prediction.................. 113
      7.1.2   Rigorous Stochastic Attractors................................. 114
      7.1.3   Skeleton Models and the Monsoon 114
   7.2   The Skeleton and "Muscle" of the MJO 114
   7.3   Other Models for the MJO.. 115
   7.4   Improving the MJO and Monsoon in GCMs Through
      Judicious Stochastic Parameterization 116
   References 117

Index.. 121

Chapter 1
Introduction

A grand challenge in contemporary climate, atmosphere, and ocean science is to understand and predict intraseasonal variability for time scales from 30 to 60 days, which is longer than standard weather time scales of at most a week and much shorter than the yearly time scales of short-term climate. In a prescient report from the 1950s, von Neumann (1960) called such problems at the intersection of weather and climate the greatest challenge in future meteorology (Moncrieff et al. 2007; Lau and Waliser 2012). It turns out that the most significant and largest intraseasonal variability occurs in the tropics through the Indian–Asian monsoon and the Madden–Julian Oscillation (MJO), named after its discoverers in the early 1970s (Madden and Julian 1971, 1972). The warmest surface waters in the world reside in the "warm pool" at the equator ranging from the Indian Ocean to the western Pacific and centered at the Indonesian Maritime Continent; this warm water is a significant source of available energy for the atmosphere.

The MJO is an equatorial wave envelope of complex multi-scale convective processes, coupled with planetary-scale ($\approx$10,000–40,000 km) circulation anomalies. Individual MJO events often begin with a standing wave in the Indian Ocean, followed by eastward propagation across the western Pacific Ocean at a speed of roughly 5 m/s (Zhang 2005). In addition to its significance in its own right, the MJO also significantly affects many other components of the atmosphere–ocean–earth system, such as monsoon development, intraseasonal predictability in mid-latitudes, and the development of the El Niño southern Oscillation (ENSO) (Zhang 2005; Lau and Waliser 2012).

Despite the widespread importance of the MJO, many computer general circulation models (GCMs) have typically had poor representations of it (Lin et al. 2006; Kim et al. 2009), although there has been significant recent progress in a few GCMs (see, e.g., Khouider et al. 2011; Ahn et al. 2017, and the references therein). Moreover, simple theories for the MJO have also struggled to reproduce all of the MJO's fundamental features together and to explain the MJO's fundamental mechanisms. Reviews of many of the MJO models that have been proposed can be

© The Author(s), under exclusive licence to Springer Nature Switzerland AG 2019
A. J. Majda et al., *Tropical Intraseasonal Variability and the Stochastic Skeleton Method*, SpringerBriefs in Mathematics of Planet Earth, https://doi.org/10.1007/978-3-030-22247-5_1

found in, e.g., Zhang (2005), Lau and Waliser (2012), and the references therein. While many models have been proposed, none has been generally accepted. The present book herein describes a summary of the authors' body of work and views on the MJO.

While theory and simulation of the MJO remain difficult challenges, observational analyses have identified the basic features of the MJO on intraseasonal/planetary scales including

(I) Slow eastward phase speed of roughly 5 m/s (Hendon and Salby 1994; Hendon and Liebmann 1994; Maloney and Hartmann 1998; Kiladis et al. 2005),

(II) Peculiar dispersion relation with $d\omega/dk \approx 0$ (Salby and Hendon 1994; Wheeler and Kiladis 1999; Roundy and Frank 2004), and

(III) Horizontal quadrupole vortex structure (Hendon and Salby 1994; Hendon and Liebmann 1994; Maloney and Hartmann 1998).

These features are on large scales and are referred to here as the MJO's "skeleton" in order to contrast with additional details of the MJO which we refer to as the MJO's "muscle." Furthermore, recent observations (Yoneyama et al. 2013) also show that

(IV) MJOs often occur intermittently in time and are

(V) organized into MJO wave trains with growth and demise.

Individual MJOs sometimes circumnavigate the entire equator, while others grow locally, propagate thousands of kilometers, and then terminate. MJOs have characteristic large-scale patterns of winds, pressure, temperature, water vapor, and precipitation with large societal impacts.

The goal of this monograph is to show how one can use modern applied mathematics and physical insight to construct the simplest and first nonlinear dynamical model for the MJO, the *(MJO) Stochastic Skeleton Model*, which simultaneously captures all the features (I)–(V) of the MJO (Majda and Stechmann 2009, 2011; Thual et al. 2014). Since the (MJO) Skeleton Model offers a theoretical prediction of the MJO structure, this leads to new detailed methods to identify the MJO in observational data both to provide new physical insight and to validate the skill of the Skeleton Model itself (Stechmann and Majda 2015; Stechmann and Ogrosky 2014; Stachnik et al. 2015).

The modern applied math modus operandi utilized here blends linear and nonlinear partial differential equations (PDEs), simple stochastic modeling, and numerical algorithms together with physical insight. The monograph is largely self-contained with a few basic math references given here as general background (Majda 2003; Khouider et al. 2013; Lawler 1995; Majda et al. 2008). The main text is the five Chaps. 2–6 with detailed discussion and references for each chapter. Chapter 2 introduces and studies the deterministic nonlinear skeleton model, which captures all the features in (I)–(III). Chapter 3 develops and motivates the Stochastic Skeleton Model, which simultaneously predicts all five observed features (I)–(V). Chapter 5 develops comparisons of the detailed theoretical predictions with observational data. These three chapters form the basic core of the monograph. Chapters 4 and 6 treat other important topics including skeleton models for the

MJO with refined vertical structure and simplified models for interaction of the MJO and the extratropics. The monograph ends with a brief discussion in Chap. 7 of contemporary and future research directions in this exciting and important multi-disciplinary research topic.

References

Ahn MS, Kim D, Sperber KR, Kang IS, Maloney E, Waliser D, Hendon H et al (2017) MJO simulation in CMIP5 climate models: MJO skill metrics and process-oriented diagnosis. Clim Dyn 49(11–12):4023–4045

Hendon HH, Liebmann B (1994) Organization of convection within the Madden–Julian oscillation. J Geophys Res 99:8073–8084. https://doi.org/10.1029/94JD00045

Hendon HH, Salby ML (1994) The life cycle of the Madden–Julian oscillation. J Atmos Sci 51:2225–2237

Khouider B, St-Cyr A, Majda AJ, Tribbia J (2011) The MJO and convectively coupled waves in a coarse-resolution GCM with a simple multicloud parameterization. J Atmos Sci 68:240–264

Khouider B, Majda AJ, Stechmann SN (2013) Climate science in the tropics: waves, vortices and PDEs. Nonlinearity 26(1):R1–R68

Kiladis GN, Straub KH, Haertel PT (2005) Zonal and vertical structure of the Madden–Julian oscillation. J Atmos Sci 62:2790–2809

Kim D, Sperber K, Stern W, Waliser D, Kang IS, Maloney E, Wang W, Weickmann K, Benedict J, Khairoutdinov M, et al (2009) Application of MJO simulation diagnostics to climate models. J Clim 22(23):6413–6436

Lau WKM, Waliser DE (eds) (2012) Intraseasonal variability in the atmosphere–ocean climate system, 2nd edn. Springer, Berlin

Lawler G (1995) Introduction to stochastic processes. Chapman & Hall/CRC, Boca Raton

Lin JL, Kiladis GN, Mapes BE, Weickmann KM, Sperber KR, Lin W, Wheeler M, Schubert SD, Del Genio A, Donner LJ, Emori S, Gueremy JF, Hourdin F, Rasch PJ, Roeckner E, Scinocca JF (2006) Tropical intraseasonal variability in 14 IPCC AR4 climate models Part I: Convective signals. J Clim 19:2665–2690

Madden RA, Julian PR (1971) Detection of a 40–50 day oscillation in the zonal wind in the tropical Pacific. J Atmos Sci 28(5):702–708

Madden RA, Julian PR (1972) Description of global-scale circulation cells in the Tropics with a 40–50 day period. J Atmos Sci 29:1109–1123

Majda AJ (2003) Introduction to PDEs and waves for the atmosphere and ocean. Courant lecture notes in mathematics, vol 9. American Mathematical Society, Providence

Majda AJ, Stechmann SN (2009) The skeleton of tropical intraseasonal oscillations. Proc Natl Acad Sci USA 106(21):8417–8422

Majda AJ, Stechmann SN (2011) Nonlinear dynamics and regional variations in the MJO skeleton. J Atmos Sci 68:3053–3071

Majda AJ, Franzke C, Khouider B (2008) An applied mathematics perspective on stochastic modeling for climate. Philos Trans R Soc A 366:2427–2453

Maloney ED, Hartmann DL (1998) Frictional moisture convergence in a composite life cycle of the Madden–Julian oscillation. J Clim 11:2387–2403

Moncrieff MW, Shapiro M, Slingo J, Molteni F (2007) Collaborative research at the intersection of weather and climate. WMO Bull 56:204–211

Roundy P, Frank W (2004) A climatology of waves in the equatorial region. J Atmos Sci 61(17):2105–2132

Salby M, Hendon H (1994) Intraseasonal behavior of clouds, temperature, and motion in the tropics. J Atmos Sci 51(15):2207–2224

Stachnik JP, Waliser DE, Majda AJ, Stechmann SN, Thual S (2015) Evaluating MJO event initiation and decay in the skeleton model using an RMM-like index. J Geophys Res Atmos 120(22). https://doi.org/10.1002/2015JD023916

Stechmann SN, Majda AJ (2015) Identifying the skeleton of the Madden–Julian oscillation in observational data. Mon Weather Rev 143:395–416. https://doi.org/10.1175/MWR-D-14-00169.1

Stechmann SN, Ogrosky HR (2014) The Walker circulation, diabatic heating, and outgoing longwave radiation. Geophys Res Lett 41:9097–9105. https://doi.org/10.1002/2014GL062257

Thual S, Majda AJ, Stechmann SN (2014) A stochastic skeleton model for the MJO. J Atmos Sci 71:697–715

von Neumann J (1960) Some remarks on the problem of forecasting climatic fluctuations. In: Dynamics of climate. Elsevier, Amsterdam, pp 9–11

Wheeler M, Kiladis GN (1999) Convectively coupled equatorial waves: analysis of clouds and temperature in the wavenumber–frequency domain. J Atmos Sci 56(3):374–399

Yoneyama K, Zhang C, Long CN (2013) Tracking pulses of the Madden-Julian oscillation. Bull Am Meteor Soc 94(12):1871–1891

Zhang C (2005) Madden–Julian Oscillation. Rev Geophys 43:RG2003. https://doi.org/10.1029/2004RG000158

Chapter 2
The Deterministic Skeleton Model and Observed Features of the MJO

This chapter describes the deterministic version of the skeleton model of Majda and Stechmann (2009b), a model for tropical intraseasonal variability. The dominant component of intraseasonal ($\approx$30–60 days) variability in the tropics is called the Madden–Julian Oscillation (MJO), which was first discovered in the 1970s and has remained perplexing ever since. In this chapter, the basic observed features of the MJO are described, and they form the basis for the description of the physical mechanisms proposed in the skeleton model. Both linear and nonlinear versions of the deterministic skeleton model are described in this chapter. The deterministic version of the skeleton model also forms the foundation for the subsequent chapters which incorporate additional features such as, for instance, stochastic mechanisms to account for the irregular dynamics of the MJO (Chap. 3) and multiscale interactions with the extratropics to investigate the global impact of the MJO (Chap. 4).

2.1 Introduction

The Madden–Julian Oscillation (MJO) is the dominant component of intraseasonal ($\approx$30–60 days) variability in the tropics (Madden and Julian 1971, 1972, 1994). It is an equatorial wave envelope of complex multi-scale convective processes, coupled with planetary-scale ($\approx$10,000–40,000 km) circulation anomalies. Individual MJO events often begin over the Indian Ocean and propagate eastward across the western Pacific Ocean at a speed of roughly 5 m/s (Zhang 2005), much slower than typical gravity wave speeds.

In addition to its significance in its own right, the MJO also significantly affects many other components of the atmosphere–ocean–earth system, such as monsoon development, intraseasonal predictability in mid-latitudes, and the development of the El Niño southern oscillation (ENSO) (Lau and Waliser 2012; Zhang 2005).

© The Author(s), under exclusive licence to Springer Nature Switzerland AG 2019
A. J. Majda et al., *Tropical Intraseasonal Variability and the Stochastic Skeleton Method*, SpringerBriefs in Mathematics of Planet Earth, https://doi.org/10.1007/978-3-030-22247-5_2

Despite the widespread importance of the MJO, many computer general circulation models (GCMs) have typically had poor representations of it (Lin et al. 2006; Kim et al. 2009), although there has been significant recent progress in a few GCMs (see, e.g., Khouider et al. 2011; Ahn et al. 2017, and the references therein). Furthermore, MJO predictions are believed to be far from reaching their potential upper limits of predictability (Waliser 2012). Such challenges indicate a need for improved understanding of the MJO through simplified models.

In this chapter, a simplified model is presented for the MJO. The model is called the MJO skeleton model and was originally described by Majda and Stechmann (2009b). The model is able to reproduce many of the MJO's fundamental features together, and it offers a mechanism for the MJO that can help improve conceptual understanding and simulation of the MJO.

The chapter is organized as follows. The observed features of the MJO are reviewed in Sect. 2.2, the MJO skeleton model is described in Sect. 2.3, and linear and nonlinear model results are presented in Sects. 2.4, 2.5, and 2.6.

2.2 Observed Features of the MJO

While theory and simulation of the MJO remain difficult challenges, observational analyses have identified the basic features of the MJO on intraseasonal/planetary scales including

I. Slow eastward phase speed of roughly 5 m/s (Hendon and Salby 1994; Hendon and Liebmann 1994; Maloney and Hartmann 1998; Kiladis et al. 2005),

II. Peculiar dispersion relation with $d\omega/dk \approx 0$ (Salby and Hendon 1994; Wheeler and Kiladis 1999; Roundy and Frank 2004), and

III. Horizontal quadrupole vortex structure (Hendon and Salby 1994; Hendon and Liebmann 1994; Maloney and Hartmann 1998).

Figure 2.1 displays several MJO events with slow eastward phase speed of roughly 5 m/s, Fig. 2.2 indicates the peculiar dispersion relation with $d\omega/dk \approx 0$, and Fig. 2.3 shows the MJO's horizontal quadrupole vortex structure. These features are on large scales and are referred to here as the MJO's "skeleton" in order to contrast with additional details of the MJO which we refer to as the MJO's "muscle."

In addition to the salient planetary/intraseasonal features of MJO composites, individual MJO events often have additional features, such as westerly wind bursts (Lin and Johnson 1996; Majda and Biello 2004; Biello and Majda 2005; Majda and Stechmann 2009a), complex vertical structures (Lin and Johnson 1996; Myers and Waliser 2003; Kikuchi and Takayabu 2004; Kiladis et al. 2005; Tian et al. 2006), and complex convective features within the MJO envelope (Nakazawa 1988; Hendon and Liebmann 1994; Dunkerton and Crum 1995; Yanai et al. 2000; Houze et al. 2000; Masunaga et al. 2006; Kiladis et al. 2009). For example, smaller-scale convective features can be seen within the MJO envelopes in Fig. 2.1. Since these additional features add detailed character to each MJO's structure, and since these

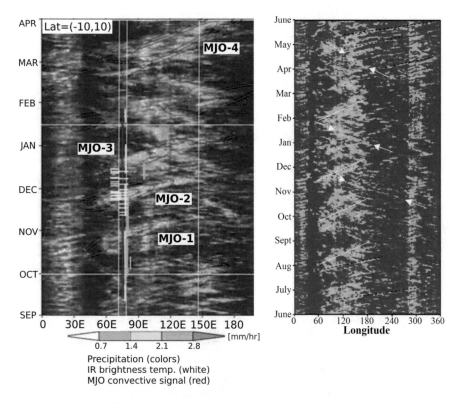

Fig. 2.1 Examples of the evolution of tropical rainfall, averaged from 10°S to 10°N, as a function of longitude and time. Green color indicates stronger rainfall, and blue color indicates weaker rainfall. Left: from the years 2011–2012, showing only the eastern hemisphere; from Yoneyama et al. (2013), ©American Meteorological Society; used with permission. Right: from the years 2000–2001, showing both eastern and western hemispheres; republished with permission from the American Geophysical Union; from Zhang (2005)

features often account for additional strength beyond the MJO's skeleton, they are referred to here as the MJO's "muscle" (Majda and Stechmann 2009b).

Recently, Majda and Stechmann (2009b) introduced a minimal dynamical model that captures the MJO's intraseasonal/planetary scale features I–III, together, for the first time in a simple model. The model is a nonlinear oscillator model for the MJO skeleton and the skeleton of tropical intraseasonal variability in general. The fundamental mechanism of the model involves interactions between (1) planetary-scale, lower-tropospheric moisture anomalies and (2) sub-planetary-scale, convection/wave activity (or, more precisely, the planetary-scale envelope of the sub-planetary-scale convective activity).

The observational relationship between (1) and (2), lower-tropospheric moisture and convective activity, is illustrated in Fig. 2.4. Lower tropospheric moisture is seen to precede the maximum in the MJO's convective activity, and this will be used as an important component of the MJO skeleton model, as described next.

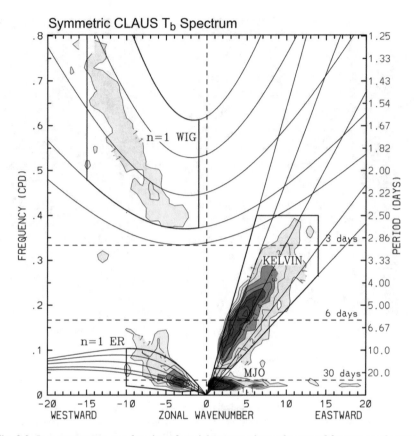

Fig. 2.2 Power spectrum, as a function of spatial wavenumber and temporal frequency, of a proxy for tropical cloudiness in observational data. A smoothed background spectrum was removed as in Wheeler and Kiladis (1999). Republished with permission from the American Geophysical Union; from Kiladis et al. (2009)

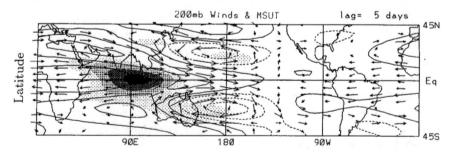

Fig. 2.3 Composite MJO structure from observational data. Dark shading indicates enhanced convection. From Hendon and Salby (1994), ©American Meteorological Society; used with permission

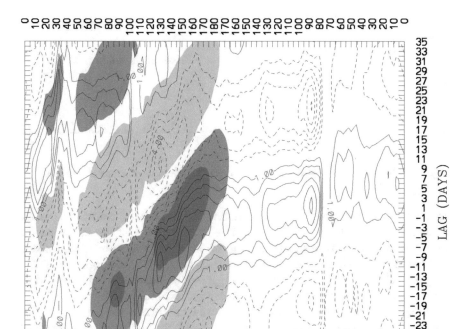

Fig. 2.4 Observations of lower-tropospheric water vapor (contours), which precede the maximum in MJO convection (dark shading). From Kiladis et al. (2005), ©American Meteorological Society; used with permission

2.3   The MJO Skeleton Model

2.3.1   Model Description

The MJO skeleton model was originally proposed and developed by Majda and Stechmann (2009b). It is a *nonlinear oscillator* model for the MJO skeleton as a neutrally stable wave; i.e., the model includes neither damping nor instability mechanisms. The fundamental mechanism of the oscillation involves interactions between (1) planetary-scale, lower-tropospheric moisture anomalies and (2) sub-planetary-scale, convection/wave activity (or, more precisely, the planetary-scale envelope of the sub-planetary-scale convective activity; see Fig. 2.5). These quantities are represented by the variables q and a, respectively:

q : lower tropospheric moisture anomaly,

a : amplitude of the convection/wave activity envelope.

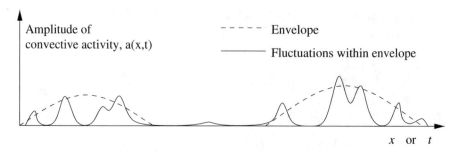

Fig. 2.5 A large-scale envelope of convective activity with fluctuations embedded within it

It is noteworthy that, for the MJO skeleton model, it is only the amplitude of the convection/wave activity envelope that is needed, not any of the details of the particular convection/waves that make up the envelope (Nakazawa 1988; Hendon and Liebmann 1994; Yanai et al. 2000; Houze et al. 2000; Masunaga et al. 2006; Kiladis et al. 2009; Dias et al. 2013), although the specific details can be important for convective momentum transports or other features of the MJO's "muscle."

A key part of the q-and-a interaction is how the moisture anomalies influence the convection. The premise is that, for convective activity on planetary/intraseasonal scales, it is the *time tendency* of convective activity—not the convective activity itself—that is most directly related to the (lower-tropospheric) moisture anomaly. In other words, rather than a functional relationship $a = a(q)$, it is posited that q mainly influences the tendency—i.e., the growth and decay rates—of the convective activity. The simplest equation that embodies this idea is

$$a_t = \Gamma q a, \tag{2.1}$$

where Γ is a constant of proportionality: positive (negative) low-level moisture anomalies create a tendency to enhance (decrease) the envelope of convection/wave activity.

The basis for (2.1), and the physics behind it, comes from a combination of observations, modeling, and theory. Generally speaking, it is well-known that tropospheric moisture content plays a key role in regulating convection (Austin 1948; Malkus 1954; Brown and Zhang 1997; Tompkins 2001; Derbyshire et al. 2004; Grabowski and Moncrieff 2004; Holloway and Neelin 2009; Waite and Khouider 2010). In observations specifically on planetary/intraseasonal scales, several studies have shown that the lower troposphere tends to moisten during the suppressed convection phase of the MJO, and lower tropospheric moisture leads the MJO's heating anomaly (Fig. 2.4 of the present chapter; Myers and Waliser 2003; Kikuchi and Takayabu 2004; Kiladis et al. 2005; Tian et al. 2006), which suggests the relationship (2.1). Furthermore, this relationship is also suggested by simplified models for synoptic-scale convectively coupled waves (Khouider and Majda 2006, 2008; Majda and Stechmann 2009a; Stechmann et al. 2013). These models show that the growth rates of the convectively coupled waves depend on the wave's environment, such as the environmental moisture content; and Stechmann et al. (2013) estimate the value of Γ from these growth rate variations. Lastly, amplitude equations such as (2.1) have been used in other areas of science and engineering, and they can sometimes be derived from the governing equations using systematic asymptotics [see Bourlioux and Majda (1995) and the references therein]. In the atmospheric sciences, Stechmann et al. (2013) show that amplitude equations can be used as a simple model for convectively coupled wave–mean flow interactions (Majda and Stechmann 2009a).

By combining the parameterization (2.1) with the (long-wave scaled) linearized primitive equations, the skeleton model of Majda and Stechmann (2009b) is obtained:

$$u_t - yv = -p_x \tag{2.2a}$$

$$yu = -p_y \tag{2.2b}$$

$$0 = -p_z + \theta \tag{2.2c}$$

$$u_x + v_y + w_z = 0 \tag{2.2d}$$

$$\theta_t + w = \bar{H}a - s^\theta \tag{2.2e}$$

$$q_t - \tilde{Q}w = -\bar{H}a + s^q \tag{2.2f}$$

$$a_t = \Gamma qa. \tag{2.2g}$$

Here u, v, and w are the zonal, meridional, and vertical velocity anomalies, respectively; p and θ are the pressure and potential temperature anomalies, respectively; and s^θ and s^q are sources of cooling and moistening, respectively. The convective heating and drying are taken to be proportional to the envelope of convection/wave activity: $\bar{H}a$. Equatorial long-wave scaling has been used (Majda 2003), and the equations have been nondimensionalized in standard fashion (Majda and Stechmann 2009a).

Notice that this model contains a minimal number of parameters: $\tilde{Q} = 0.9$, the (nondimensional) mean background vertical moisture gradient; and $\Gamma = 1.66$, or $\Gamma \approx 0.3 \text{ K}^{-1} \text{ d}^{-1}$ in dimensional units. These were the standard parameter values used by Majda and Stechmann (2009b). The source terms s^θ and s^q must also be specified (see below). Also notice that the parameter $\bar{H}$ is actually irrelevant to the dynamics (as can be seen by rescaling a); it is written here for clarity of presentation: dimensionally, it gives $\bar{H}a$ the units of a heating rate while keeping a nondimensional. The dimensional value of $\bar{H}$ was chosen to be 10 K/day so that a typical value of a is ≈ 0.1, similar to the nondimensional value of u.

To obtain the simplest model for the MJO, truncated vertical and meridional structures are used. We first describe an overview of the derivation; further details are described below in Sect. 2.3.3. For the vertical truncation, only the first baroclinic mode is used so that $u(x, y, z, t) = u(x, y, t)\sqrt{2}\cos(z)$, etc., with a slight abuse of notation. The resulting equations resemble a time-dependent version of a Matsuno–Gill model (Matsuno 1966; Gill 1980), without damping, plus equations for q and a:

$$u_t - yv - \theta_x = 0 \qquad (2.3a)$$

$$yu - \theta_y = 0 \qquad (2.3b)$$

$$\theta_t - u_x - v_y = \bar{H}a - s^\theta \qquad (2.3c)$$

$$q_t + \tilde{Q}(u_x + v_y) = -\bar{H}a + s^q \qquad (2.3d)$$

$$a_t = \Gamma qa. \qquad (2.3e)$$

Next, for the meridional truncation, it is assumed that a, the envelope of convection/wave activity, has a simple equatorial meridional structure proportional to $\exp(-y^2/2)$: $a(x, y, t) = [\bar{A}(x) + A(x, t)]\pi^{-1/4}\exp(-y^2/2)$, where $\bar{A}(x)$ is a background state. For the long-wave-scaled equations, such a meridional heating structure is known to excite only Kelvin waves and the first symmetric equatorial Rossby waves (Majda 2003; Biello and Majda 2006), and the resulting meridionally truncated equations can be written as

$$K_t + K_x = -\frac{1}{\sqrt{2}}\bar{H}A \qquad (2.4a)$$

$$R_t - \frac{1}{3}R_x = -\frac{2\sqrt{2}}{3}\bar{H}A \qquad (2.4b)$$

$$Q_t + \frac{1}{\sqrt{2}}\tilde{Q}K_x - \frac{1}{6\sqrt{2}}\tilde{Q}R_x = \left(-1 + \frac{1}{6}\tilde{Q}\right)\bar{H}A \qquad (2.4c)$$

$$A_t = \Gamma Q(\bar{A} + A), \qquad (2.4d)$$

where K and R are the amplitudes of the Kelvin and equatorial Rossby waves, respectively, and they have the associated meridional structures as shown in Fig. 1 of Majda and Stechmann (2009b).

An important point is that $K(x, t)$ and $R(x, t)$ are the amplitudes of the *structures* of Kelvin and Rossby waves, but these amplitudes in (2.4) need not always *propagate* like "dry" waves. In the absence of forcing in (2.4), the "dry" long-wave Kelvin and equatorial Rossby wave solutions are dispersionless waves that propagate at 50 and 17 m/s, respectively (Majda 2003; Biello and Majda 2006). However, in the presence of the coupled dynamical forcing A in (2.4), the Kelvin and equatorial Rossby wave *structures* can be coupled to each other and to Q and A; and these coupled modes/structures can have propagation speeds very different from 50 or 17 m/s, and they can be dispersive. One such mode has the structure and dispersion characteristics of the MJO, as shown by Majda and Stechmann (2009b) and summarized below.

The variables u, v, θ are recovered by using the formulas (Majda 2003; Biello and Majda 2006)

$$u(x, y) = \frac{1}{\sqrt{2}} \left[K(x) - \frac{1}{2} R(x) \right] \phi_0(y) + \frac{1}{4} R(x) \phi_2(y) \tag{2.5a}$$

$$v(x, y) = \left[\frac{1}{3} \partial_x R(x) - \frac{1}{3\sqrt{2}} \bar{H} A(x) \right] \phi_1(y), \tag{2.5b}$$

$$\theta(x, y) = -\frac{1}{\sqrt{2}} \left[K(x) + \frac{1}{2} R(x) \right] \phi_0(y) - \frac{1}{4} R(x) \phi_2(y) \tag{2.5c}$$

where $\phi_0(y) = \pi^{-1/4} \exp(-y^2/2)$, $\phi_1(y) = \pi^{-1/4} \sqrt{2} y \exp(-y^2/2)$, and $\phi_2(y) = \pi^{-1/4} 2^{-1/2} (2y^2 - 1) \exp(-y^2/2)$ are parabolic cylinder functions (Majda 2003; Biello and Majda 2006). The meridional structures of q and the source terms are given by $q(x, y, t) = Q(x, t) \phi_0(y)$, $s^\theta(x, y) = S^\theta(x) \phi_0(y)$, and $s^q(x, y) = S^q(x) \phi_0(y)$. In the simplest case, $S^\theta = \overline{S^\theta} = 1$ K/day is the uniform radiative cooling rate. Also notice that $A(x, t)$ is an anomaly from a base state $\bar{A}(x)$, which is chosen to balance the sources of cooling and moistening: $\bar{H} \bar{A}(x) = S^\theta(x) = S^q(x)$. In this fashion, $\bar{A}(x)$ represents base state variations in the simplest way, and $\bar{A}(x)$ will be used to represent the sea surface temperature (SST), as either a uniform SST that is independent of x or a warm pool SST with zonal variations.

2.3.2 Energetics

The nonlinear MJO skeleton model has two important energy principles, which both hold even in the presence of source terms $s^\theta(x)$ and $s^q(x)$ provided that $s^\theta(x) = s^q(x)$. First, the model (2.3) conserves a vertically integrated moist static energy:

$$\partial_t(\theta + q) - (1 - \tilde{Q})(u_x + v_y) = 0. \tag{2.6}$$

Second, the model (2.3) conserves a positive total energy that includes a contribution from the convective activity a:

$$\partial_t \left[\frac{1}{2} u^2 + \frac{1}{2} \theta^2 + \frac{1}{2} \frac{\tilde{Q}}{1 - \tilde{Q}} \left(\theta + \frac{q}{\tilde{Q}} \right)^2 + \frac{\bar{H}}{\Gamma \tilde{Q}} a - \frac{s}{\Gamma \tilde{Q}} \log a \right] - \partial_x(u\theta) - \partial_y(v\theta) = 0, \tag{2.7}$$

where $s = s^\theta = s^q$. This total energy is a sum of four terms: dry kinetic energy $u^2/2$, dry potential energy $\theta^2/2$, a moist potential energy proportional to $(\theta + \tilde{Q}^{-1} q)^2$ [cf. Frierson et al. (2004)], and a convective energy $(\bar{H}/\Gamma \tilde{Q})a - (s/\Gamma \tilde{Q}) \log a$. Note that the natural requirement on the background moisture gradient, $0 < \tilde{Q} < 1$, is needed to guarantee a positive energy. Also, note that this energy is a convex function of u, θ, q, and a.

The meridionally truncated system in (2.4) possesses similar conservation laws (Chen and Stechmann 2016). For example, the energy conservation principle is

$$\partial_t \left[\frac{1}{2}K^2 + \frac{3}{16}R^2 + \frac{1}{2}\frac{\tilde{Q}}{1 - \tilde{Q}}\left(\frac{Q}{\tilde{Q}} - \frac{K}{\sqrt{2}} - \frac{R}{2\sqrt{2}}\right)^2 + \frac{\bar{H}}{\Gamma\tilde{Q}}A - \frac{S}{\Gamma\tilde{Q}}\log A \right]$$

$$- \partial_x \left(\frac{1}{2}K^2 - \frac{1}{16}R^2 \right) = 0, \tag{2.8}$$

where $S(x) = S^\theta(x) = S^q(x)$.

2.3.3 Derivation of K, R, Q, and A Variables and Evolution Equations

In this subsection, some selected aspects are presented for the derivations of (2.4) from (2.2), including the derivations of the variables K and R. A main part of this derivation is the progression from primitive variables (u, θ) to characteristic variables (r, l) to equatorial wave variables $(K, R_1, R_2, \cdots)$. For more information, see Gill (1980, 1982), Majda (2003), Biello and Majda (2006), Majda and Stechmann (2009b), and Stechmann and Majda (2015).

2.3.3.1 Meridional Basis Functions

For equatorial waves, convenient meridional basis functions are the parabolic cylinder functions:

$$\phi_m(y) = \frac{1}{(m!\sqrt{\pi}\,2^m)^{1/2}} H_m(y)\,e^{-y^2/2}, \quad m = 0, 1, 2, \cdots \tag{2.9}$$

where $H_m(y)$ are the Hermite polynomials:

$$H_m(y) = (-1)^m e^{+y^2} \frac{d^m}{dy^m} e^{-y^2}. \tag{2.10}$$

The first few parabolic cylinder functions are

$$\phi_0(y) = \pi^{-1/4} \exp(-y^2/2), \tag{2.11a}$$

$$\phi_1(y) = \pi^{-1/4}\sqrt{2}\,y\,\exp(-y^2/2), \tag{2.11b}$$

$$\phi_2(y) = \pi^{-1/4}2^{-1/2}(2y^2 - 1)\exp(-y^2/2). \tag{2.11c}$$

The functions $\phi_m(y)$ form an orthonormal basis, and any variable such as $u(x, y, t)$ can then be expanded as

$$u(x, y, t) = \sum_{m=0}^{\infty} u_m(x, t)\, \phi_m(y) \tag{2.12}$$

where the quantities $u_m(x, t)$ are obtained using the projection

$$u_m(x, t) = \int_{-\infty}^{\infty} u(x, y, t)\phi_m(y)\, dy \tag{2.13}$$

Formulas analogous to (2.12)–(2.13) also apply to v, θ, q, a.

2.3.3.2   Characteristic Variables l and r

The equatorial long-wave equations can be written naturally in terms of characteristic variables r and l:

$$r = \frac{1}{\sqrt{2}}(u - \theta), \qquad l = \frac{1}{\sqrt{2}}(u + \theta), \tag{2.14}$$

$$f^r = \frac{1}{\sqrt{2}}(f^u - f^\theta) = -\frac{1}{\sqrt{2}}(\bar{H}a - s^\theta), \tag{2.15}$$

$$f^l = \frac{1}{\sqrt{2}}(f^u + f^\theta) = +\frac{1}{\sqrt{2}}(\bar{H}a - s^\theta), \tag{2.16}$$

where f^u and f^θ are the forcing terms (right-hand sides) of the equations for u and θ, (2.3a) and (2.3c), respectively. Specifically, here in (2.3a) and (2.3c), we have $f^u = 0$ and $f^\theta = \bar{H}a - s^\theta$. In terms of these characteristics variables, (2.3a)–(2.3c) can be rewritten as

$$r_t + r_x + L_- v = f^r \tag{2.17a}$$

$$l_t - l_x - L_+ v = f^l \tag{2.17b}$$

$$L_+ r - L_- l = 0 \tag{2.17c}$$

where L_- and L_+ are the raising and lowering operators, respectively:

$$L_\pm = \frac{1}{\sqrt{2}}(\partial_y \pm y), \tag{2.18}$$

which operate on parabolic cylinder functions as

$$L_+\phi_m = \sqrt{m}\phi_{m-1}, \qquad L_-\phi_m = -\sqrt{m+1}\phi_{m+1} \tag{2.19}$$

Note that the meridional velocity equation has become (2.17c), which describes meridional geostrophic balance in terms of the characteristic variables.

2.3.3.3 Defining K, R_1, R_2, R_3, $\cdots$

Next, we pass from characteristic variables (r, l) to the equatorial wave variables $(K, R_1, R_2, R_3, \cdots)$. This change of variables is facilitated by an expansion in meridional basis functions $\phi_m(y)$, similar to (2.9)–(2.13):

$$\begin{pmatrix} r \\ l \\ v \end{pmatrix} = \begin{pmatrix} r_0\phi_0 \\ 0 \\ 0 \end{pmatrix} + \begin{pmatrix} 0 \\ 0 \\ v_0\phi_0 \end{pmatrix} + \sum_{m=1}^{\infty} \begin{pmatrix} r_m\phi_{m+1} \\ l_m\phi_{m-1} \\ v_m\phi_m \end{pmatrix} \tag{2.20}$$

Each term on the right-hand side corresponds to a different wave type. By combining this expansion with (2.17a)–(2.17c), the evolution equation of each wave type can be obtained as follows.

The first term in the expansion (2.20) corresponds to the Kelvin wave, which from (2.17a) evolves as

$$\partial_t r_0 + \partial_x r_0 = f_0^r = -\frac{1}{\sqrt{2}}(\bar{H}a_0 - s_0^\theta) \tag{2.21}$$

Replacing the symbol r_0 with K leads to (2.4a).

The second term in the expansion (2.20) corresponds to the mixed Rossby–gravity (MRG) wave. With the equatorial long-wave scaling used here, the remnant of the MRG wave is not a dynamical equation but a diagnostic relation: $v_0 = -f_1^r = +(\bar{H}a_1 - s_1^\theta)/\sqrt{2}$, which follows from (2.17a). Since the antisymmetric convective activity a_1 is not included in the simplest skeleton model with only K and R, the MRG remnant plays no role here.

Finally, in the infinite sum in (2.20), each m corresponds to the mth Rossby wave. The sum of all types of Rossby waves can be written as an expansion in parabolic cylinder functions as

$$R_{\text{total}} = L_+r + L_-l = \sum_{m=1}^{\infty} R_m\phi_m \tag{2.22}$$

To get an equation for R_{total}, apply L_+ to (2.17a), apply L_- to (2.17b), and then add or subtract to get, respectively,

$$\partial_t R_{\text{total}} + (L_+L_- - L_-L_+)v = L_+ f^r + L_- f^l \tag{2.23}$$

$$\partial_x R_{\text{total}} + (L_+L_- + L_-L_+)v = L_+ f^r - L_- f^l \tag{2.24}$$

These equations can be written in terms of meridional modes as

$$\partial_t R_m - v_m = \sqrt{m+1} f^r_{m+1} - \sqrt{m} f^l_{m-1} \tag{2.25}$$

$$\partial_x R_m - (2m+1)v_m = \sqrt{m+1} f^r_{m+1} + \sqrt{m} f^l_{m-1} \tag{2.26}$$

To get an equation in terms of R_m alone, v_m can be eliminated to yield

$$\partial_t R_m - \frac{1}{2m+1} \partial_x R_m = \frac{2m\sqrt{m+1}}{2m+1} f^r_{m+1} - \frac{(2m+2)\sqrt{m}}{2m+1} f^l_{m-1} \tag{2.27}$$

In the simplest version of the skeleton model, only a_0 is retained and hence only R_1 is excited. Taking (2.27) for $m = 1$ and replacing the symbol R_1 with R then leads to (2.4b).

2.3.3.4   Recovering u and θ from K, R_1, R_2, R_3, $\cdots$

The variables u, v, θ are recovered in a two-step process. First, from K and R_m, $m = 1, 2, 3, \cdots$, one can recover the characteristic variables l and r using meridional geostrophic balance (2.17c) ($\sqrt{m+1} r_m + \sqrt{m} l_m = 0$) and the definition $R_m = \sqrt{m+1} r_m - \sqrt{m} l_m$ from (2.22). Second, from r and l, one can recover u and θ using (2.14). With the meridional mode truncations of this paper described above, the resulting formulas (Majda 2003; Biello and Majda 2006) were reported above in (2.5). Finally, v is recovered from (2.26), and where the standard meridional mode truncations of this paper were applied as described above.

2.3.3.5   Moisture Dynamics

The moisture $q(x, y, t)$ can be expanded as $q(x, y, t) = \sum_{m=0}^{\infty} q_m(x, t)\phi_m(y)$, where the sum is truncated here at $q_0(x, t)$, which is relabeled as $Q(x, t)$ in (2.4).

The dynamics of $Q(x, t)$ in (2.4c) can be derived from the dynamics of $q(x, y, t)$ in (2.3d) in the following way. Each term in (2.3d) can be expanded in parabolic cylinder functions and then projected onto ϕ_0 in order to find the evolution equation for $q_0(x, t)$. First, u_x can be expanded in parabolic cylinder functions using (2.5a). Next, to handle the v_y term, the operator ∂_y can be written in terms of the ladder operators as $\partial_y = (L_+ + L_-)/\sqrt{2}$, and v_m can be written in terms of R_m and the convective activity using (2.26) or (2.5b). Finally, a projection onto ϕ_0 leads to (2.4d).

2.3.3.6 Convective Activity Dynamics

The convective activity $a(x, y, t)$ can be expanded as $a(x, y, t) = \sum_{m=0}^{\infty} a_m(x, t)$ $\phi_m(y)$, where the sum is truncated here at $a_0(x, t)$, which is relabeled as $A(x, t)$ in (2.4). The dynamics of $A(x, t)$ in (2.4d) can be derived from the dynamics of $a(x, y, t)$ in (2.3e) by expanding q and a in the basis $\{\phi_m(y)\}_{m=0}^{\infty}$ and projecting (2.3e) onto ϕ_0. Assuming that q and a are truncated at the q_0 and a_0 terms, (2.4d) is obtained. Note that the projection procedure leads to the integral $\int_{-\infty}^{+\infty} \phi_0(y) \phi_0(y) \phi_0(y) \, dt$, which takes the value $\sqrt{2/3}\pi^{-1/4}$, which for notational simplicity has been absorbed into a new definition of Γ in (2.4d) compared to the Γ in (2.3e).

2.4 Linear Theory

2.4.1 Formula for Intraseasonal Oscillation Frequency

A formula for the intraseasonal oscillation frequency ω of the MJO skeleton can be obtained by considering the even simpler case of flow above the equator. In this case, (2.3) are used, v and y are set to zero, and meridional derivatives are ignored. The result is a linear system of four equations for u, θ, q, a, and the system can be solved exactly due to the perfect east–west symmetry:

$$2\omega^2 = \Gamma \bar{R} + k^2 \pm \sqrt{(\Gamma \bar{R} + k^2)^2 - 4\Gamma \bar{R} k^2 (1 - \tilde{Q})} \qquad (2.28)$$

where k is the zonal wavenumber. A simple formula for the oscillation frequency of the low-frequency waves,

$$\omega \approx \sqrt{\Gamma \bar{R}(1 - \tilde{Q})}, \qquad (2.29)$$

can be obtained from (2.28) approximately. For the standard parameter values used here, the oscillation period corresponding to (2.29) is 45 days, in agreement with observations of the MJO (Salby and Hendon 1994; Wheeler and Kiladis 1999; Roundy and Frank 2004). Notice that this formula is independent of the wavenumber k; i.e., this model recovers the peculiar dispersion relation $d\omega/dk \approx 0$ from the observational record (Salby and Hendon 1994; Wheeler and Kiladis 1999; Roundy and Frank 2004).

It will be shown below that (2.29) holds for the eastward-propagating branch of the beta-plane model (2.4) as well.

2.4.2 Linear Waves

In this section, the linear waves of the simplest model in (2.4)–(2.5) are presented. Since (2.4) involves four dynamically coupled variables, there are four linear modes. The dispersion relation for the linear modes is shown in Fig. 2.6. (Only the two low-frequency, intraseasonal modes are shown. The other two modes are high-frequency modes and are only weakly coupled to the wave activity; they will be discussed only briefly below.) Figure 2.6 shows that eastward-propagating waves, like the MJO (Salby and Hendon 1994; Wheeler and Kiladis 1999; Roundy and Frank 2004), have the peculiar dispersion relation $d\omega/dk \approx 0$. Moreover, this dispersion relation is robust over a wide range of parameter values, and the oscillation periods spanned by these reasonable parameter values are in the range of 30–60 days, which is the observed range of the MJO's oscillation period (Salby and Hendon 1994; Wheeler and Kiladis 1999; Roundy and Frank 2004). The westward-propagating waves, on the other hand, which are plotted with positive ω and negative k, have variable ω, and their oscillation periods are seasonal, not intraseasonal, for $k = 1$ and 2. This suggests the first piece of our explanation for the observed dominance of eastward-propagating intraseasonal variability: the westward-propagating modes have seasonal oscillation periods, on which time scales other phenomena are expected to dominate over modulations of synoptic scale wave activity.

The physical structure of the wavenumber-2 MJO mode is shown in Fig. 2.6 for the standard parameter values. Horizontal quadrupole vortices are prominent, as in observations (Hendon and Salby 1994; Hendon and Liebmann 1994; Maloney and Hartmann 1998), and the maximum wave activity is colocated with the maximum in equatorial convergence. The lower tropospheric moisture leads and is in quadrature with the wave activity, which is also roughly the relationship seen in observations (Kiladis et al. 2005; Tian et al. 2006; Kikuchi and Takayabu 2004). The pressure contours clearly display the mixed Kelvin/Rossby wave structure of the wave. Equatorial high pressure anomalies are colocated with the westerly wind burst as in Kelvin waves; and they are flanked by off-equatorial low pressure anomalies and cyclonic Rossby gyres, in broad agreement with the observational record (Hendon and Salby 1994; Hendon and Liebmann 1994; Maloney and Hartmann 1998). Rectification of the vertical structure and some of the phase relationships is likely due to effects of higher vertical modes (Majda and Biello 2004; Biello and Majda 2005; Majda et al. 2007; Khouider and Majda 2007).

The relative contributions of K, R, Q, and A to these linear waves are shown in Fig. 2.6 for wavenumbers 1, 2, and 3. The MJO has significant contributions from both the Kelvin and Rossby components, whereas the westward modes are dominated by the Rossby component. In addition, the larger Q and A amplitudes suggest further explanation for eastward-propagating rather than westward-propagating intraseasonal oscillations: the eastward-propagating modes are more strongly coupled to equatorial moist convective processes.

The physical structure of the low-frequency westward-propagating mode is shown in Fig. 2.6. Its circulation is almost purely Rossby wave-like with little

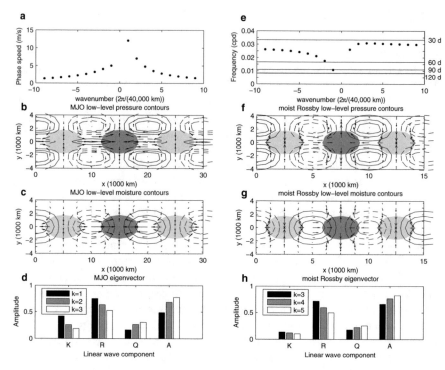

Fig. 2.6 Summary of low-frequency linear waves of the skeleton model (2.4). (**a**) Phase speed ω/k as a function of wavenumber k. Eastward (westward) propagation is denoted by positive (negative) wavenumber k. (**b**) Horizontal structure of the $k = 2$ MJO mode. Lower tropospheric velocity vectors are shown with contours of lower tropospheric pressure anomalies with positive (negative) anomalies denoted by solid (dashed) lines. The contour interval is one-fourth the maximum amplitude of the anomaly, and the zero contour is not shown. (**c**) Same as (**b**), except contours of lower tropospheric moisture anomalies. (**d**) Component amplitudes of the MJO eigenvector for wavenumbers $k = 1, 2$, and 3. (**e**) Same as (**a**), except for oscillation frequency $\omega(k)$. Horizontal lines denote oscillation periods of 30, 60, 90, and 120 days. (**f**) Same as (**b**), except for the $k = 4$ moist Rossby mode. (**g**) Same as (**c**), except for the $k = 4$ moist Rossby mode. (**h**) Same as (**d**), except for the $k = 3, 4$, and 5 moist Rossby modes. From Majda and Stechmann (2011)

character of a Kelvin wave, and the positive anomaly of low-level moisture preconditioning is confined closely to the vicinity of the equator. This wave differs from observed convectively coupled equatorial Rossby waves in several respects, such as its equatorial (as opposed to off-equatorial) heating anomaly, low frequency, and slow propagation speed (Kiladis et al. 2009; Kiladis and Wheeler 1995; Wheeler et al. 2000; Yang et al. 2007). However, the spectral filters used for the wave structures in these observational studies tend to emphasize relatively high spatiotemporal frequencies, whereas the spectral peaks tend to occur on wavenumbers 3–5 with oscillation periods of 20–60 days (Kiladis et al. 2009; Wheeler and Kiladis 1999; Roundy and Frank 2004; Yang et al. 2007). The present models do capture features of the observed westward spectral peak for these wavenumbers.

In addition to the low-frequency modes presented above, the simplest linear model in (2.4) also has two high-frequency modes (not shown). The eastward-and westward-propagating high-frequency modes propagate at 50–60 m/s and 17–30 m/s, respectively, and, relative to the low-frequency modes, they are only weakly coupled to the wave activity A. For these and other reasons, the high-frequency modes do not appear to be related to observed convectively coupled equatorial waves (Kiladis et al. 2009; Khouider and Majda 2008). They have phase speeds comparable to the dry Kelvin and equatorial Rossby waves, and they passively carry a small moisture trace.

2.5 Nonlinear Simulations

Next consider nonlinear simulations of the skeleton model (Majda and Stechmann 2011). To represent a sea surface temperature (SST) with warm-pool and cold-pool regional variations, assume the background state $\bar{H}\bar{A}(x) = S^{\theta}(x)$ is a constant plus a sinusoid, with a warm pool in the center of the domain from $x \approx 10{,}000$–30,000 km and a cold pool elsewhere. As the initial condition, a wavenumber-2 MJO eigenmode is used.

The long-term evolution of the convective activity is shown in Fig. 2.7. After an initial adjustment period, the convective activity aligns itself over the warm pool from roughly time $t \approx 2000$ and thereafter. The MJO events have prominent phases of both active and suppressed convection, and each event has its own individual characteristics in terms of strength, lifetime, regional variations, etc. Furthermore, in addition to the prominent eastward-propagating disturbances, there are instances of localized standing oscillations throughout the domain. For instance, there are standing oscillations localized near $x \approx 11{,}000$ km during the period from roughly $t = 2300$–2500 days, toward the western end of the warm pool. This is in broad agreement with the visual appearance of standing oscillations in the Indian Ocean, often at the beginning of an MJO event (Lau and Chan 1985; Zhang and Hendon 1997; Kiladis et al. 2005).

To illustrate the details of a few MJO events, the final 200 days of the simulation are also shown in detail in the right column of Fig. 2.7, in plots of both $A(x, t)$ and $Q(x, t)$. Two rectangular boxes are drawn in to identify instances of localized standing oscillations: $x = 11{,}000$–15,000 km, $t = 3400$–3470 days, and $x = 15{,}000$–19,000 km, $t = 3440$–3530 days. Localized standing oscillations are prominent again, later, in the region $x = 15{,}000$–19,000 km, $t = 3550$–3600 days (for comparison, no box added). These plots show the details of the significant variations in the MJO events, including their amplitudes, propagation, lifetimes, and/or regional extent.

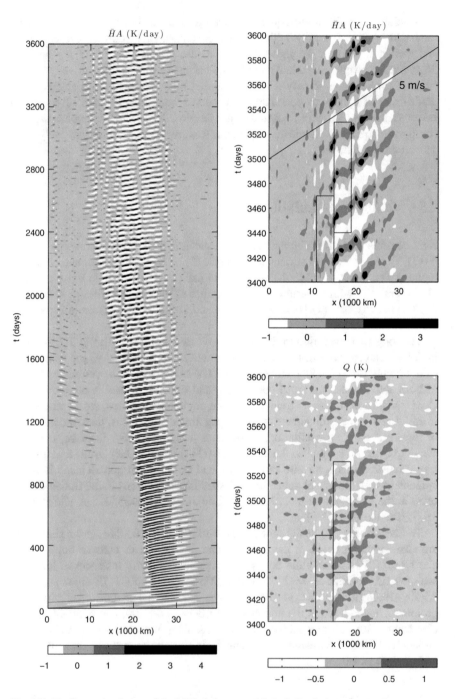

Fig. 2.7 Nonlinear simulation of the MJO skeleton model. Left: Evolution of anomalous convective activity $\bar{H}A(x, t)$ for 3600 days. Right: Zoomed-in view over the last 200 days for anomalous convective activity $\bar{H}A(x, t)$ (top) and lower tropospheric water vapor $Q(x, t)$ (bottom). From Majda and Stechmann (2011)

2.6   Nonlinear Traveling Waves: Exact Solutions

In this section, a traveling wave ansatz is applied to the system

$$K_t + K_x = -\frac{1}{2}(\bar{H}A - F) \tag{2.30a}$$

$$R_t - \frac{1}{3}R_x = -\frac{1}{3}(\bar{H}A - F) \tag{2.30b}$$

$$Q_t + \tilde{Q}K_x - \frac{\tilde{Q}}{3}R_x = \left(\frac{\tilde{Q}}{6} - 1\right)(\bar{H}A - F) \tag{2.30c}$$

$$A_t = \Gamma Q A \tag{2.30d}$$

and exact or semi-analytic solutions can be found (Chen and Stechmann 2016). (Note that these equations are equivalent to the skeleton model equations presented earlier in the chapter upon performing a rescaling of variables.) Specifically, assume that the solution has the form

$$K = K(x - st), \quad R = R(x - st), \quad Q = Q(x - st), \quad A = A(x - st). \tag{2.31}$$

Inserting this form into (2.30) leads to a simplified ODE system:

$$Q' = \frac{f(s)}{6s}(\bar{H}A - F), \tag{2.32a}$$

$$A' = -\frac{\Gamma}{s}QA, \tag{2.32b}$$

where

$$f(s) = \frac{3\tilde{Q}}{s - 1} - \frac{2\tilde{Q}}{1 + 3s} - \tilde{Q} + 6. \tag{2.33}$$

Numerical solutions of the ODE system are shown in Fig. 2.8. The waveform appears like a pulse for large energies and large amplitudes, whereas it appears like a sinusoid for small energies and small amplitudes.

One can furthermore derive an amplitude-dependent dispersion relation (see Chen and Stechmann 2016), which is also shown in Fig. 2.8. The nonlinear MJO is seen to have a lower frequency and to propagate more slowly than the linear MJO. (Note that linear waves correspond to the limit $\mathscr{A} \to 0$ of the nonlinear waves, since the linearized dynamics are derived under the assumption that the wave amplitude $\mathscr{A}$ is small.) As a specific comparison, the propagation speeds for large amplitudes ($\mathscr{A} = 0.5$) are 8.76 m/s and 4.61 m/s for wavenumbers $k = 1$ and 2, respectively, and the corresponding phase speeds for linear waves are 10.13 m/s and 5.55 m/s.

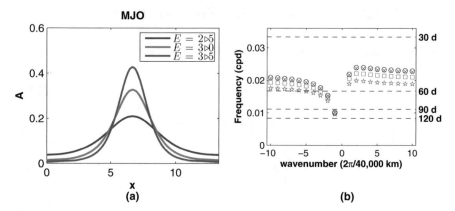

Fig. 2.8 Nonlinear traveling wave solutions to the skeleton model. (**a**) MJO waveform for convective activity, A, for different values of the total energy. (**b**) Oscillation frequency $\omega(k)$ for nonlinear and linear solutions. Different symbols correspond to different values of the wave amplitude, $\mathscr{A} = A_{max} - A_{min}$: circle, $\mathscr{A} = 0.1$; cross, $\mathscr{A} = 0.3$; square, $\mathscr{A} = 0.5$; star, $\mathscr{A} \to 0$, linear system. From Chen and Stechmann (2016)

Finally, note that a limiting case of the ODE system, (2.32), as $F \to 0$, leads to a system that can be solved analytically as

$$A(x - st) = \mathscr{A} \, \text{sech}^2 \left[\tilde{h}(s) \sqrt{\mathscr{A}} (x - st) \right].$$

This sech^2 waveform is the same as in the soliton solutions of the Korteweg–de Vries equation. The large-amplitude pulse in Fig. 2.8 is approximately of sech^2 form.

2.7 Concluding Discussion

Notice one important difference between the MJO events in the nonlinear simulation in Fig. 2.7 and in the observational data in Fig. 2.1: the MJO events in nature occur intermittently, with some periods of clusters of 1–3 successive MJO events and some periods of no MJO events. On the other hand, the nonlinear simulation in Fig. 2.7 displays a repeating sequence of one MJO event after another. This motivates the development of the stochastic skeleton model (Thual et al. 2014), which will be discussed in Chap. 3 and is able to produce intermittent generation and termination of clusters of MJO events as in the observational data in Fig. 2.1.

References

Ahn MS, Kim D, Sperber KR, Kang IS, Maloney E, Waliser D, Hendon H, et al (2017) MJO simulation in CMIP5 climate models: MJO skill metrics and process-oriented diagnosis. Clim Dyn 49(11–12):4023–4045

Austin JM (1948) A note on cumulus growth in a nonsaturated environment. J Meteorol 5(3):103–107

Biello JA, Majda AJ (2005) A new multiscale model for the Madden–Julian oscillation. J Atmos Sci 62:1694–1721

Biello JA, Majda AJ (2006) Modulating synoptic scale convective activity and boundary layer dissipation in the IPESD models of the Madden–Julian oscillation. Dyn Atmos Oceans 42:152–215

Bourlioux A, Majda AJ (1995) Theoretical and numerical structure of unstable detonations. Phil Trans R Soc Lond A 350:29–68

Brown RG, Zhang C (1997) Variability of midtropospheric moisture and its effect on cloud-top height distribution during TOGA COARE. J Atmos Sci 54(23):2760–2774

Chen S, Stechmann SN (2016) Nonlinear traveling waves for the skeleton of the Madden–Julian oscillation. Commun Math Sci 14:571–592. https://doi.org/10.4310/CMS.2016.v14.n2.a11

Derbyshire S, Beau I, Bechtold P, Grandpeix J, Piriou J, Redelsperger J, Soares P (2004) Sensitivity of moist convection to environmental humidity. Q J R Meteorol Soc 130(604):3055–3079

Dias J, Leroux S, Tulich SN, Kiladis GN (2013) How systematic is organized tropical convection within the MJO? Geophys Res Lett 40(7):1420–1425

Dunkerton TJ, Crum FX (1995) Eastward propagating ∼2- to 15-day equatorial convection and its relation to the tropical intraseasonal oscillation. J Geophys Res 100(D12):25,781–25,790

Frierson DMW, Majda AJ, Pauluis OM (2004) Large scale dynamics of precipitation fronts in the tropical atmosphere: a novel relaxation limit. Commun Math Sci 2(4):591–626

Gill AE (1980) Some simple solutions for heat-induced tropical circulation. Q J R Meteor Soc 106(449):447–462

Gill AE (1982) Atmosphere–ocean dynamics. International geophysics series, vol 30. Academic Press, New York.

Grabowski WW, Moncrieff MW (2004) Moisture-convection feedback in the Tropics. Q J R Meteorol Soc 130:3081–3104

Hendon HH, Liebmann B (1994) Organization of convection within the Madden–Julian oscillation. J Geophys Res 99:8073–8084. https://doi.org/10.1029/94JD00045

Hendon HH, Salby ML (1994) The life cycle of the Madden–Julian oscillation. J Atmos Sci 51:2225–2237

Holloway CE, Neelin JD (2009) Moisture vertical structure, column water vapor, and tropical deep convection. J Atmos Sci 66(6):1665–1683

Houze RA Jr, Chen SS, Kingsmill DE, Serra Y, Yuter SE (2000) Convection over the Pacific warm pool in relation to the atmospheric Kelvin–Rossby wave. J Atmos Sci 57:3058–3089

Khouider B, Majda AJ (2006) A simple multicloud parameterization for convectively coupled tropical waves. Part I: Linear analysis. J Atmos Sci 63:1308–1323

Khouider B, Majda AJ (2007) A simple multicloud parameterization for convectively coupled tropical waves. Part II: Nonlinear simulations. J Atmos Sci 64:381–400

Khouider B, Majda AJ (2008) Equatorial convectively coupled waves in a simple multicloud model. J Atmos Sci 65:3376–3397

Khouider B, St-Cyr A, Majda AJ, Tribbia J (2011) The MJO and convectively coupled waves in a coarse-resolution GCM with a simple multicloud parameterization. J Atmos Sci 68:240–264

Kikuchi K, Takayabu YN (2004) The development of organized convection associated with the MJO during TOGA COARE IOP: Trimodal characteristics. Geophys Res Lett 31. https://doi.org/10.1029/2004GL019601

Kiladis G, Wheeler M (1995) Horizontal and vertical structure of observed tropospheric equatorial Rossby waves. J Geophys Res 100(D11):22,981–22,998

Kiladis GN, Straub KH, Haertel PT (2005) Zonal and vertical structure of the Madden–Julian oscillation. J Atmos Sci 62:2790–2809

Kiladis GN, Wheeler MC, Haertel PT, Straub KH, Roundy PE (2009) Convectively coupled equatorial waves. Rev Geophys 47:RG2003. https://doi.org/10.1029/2008RG000266

Kim D, Sperber K, Stern W, Waliser D, Kang IS, Maloney E, Wang W, Weickmann K, Benedict J, Khairoutdinov M, et al (2009) Application of MJO simulation diagnostics to climate models. J Climate 22(23):6413–6436

Lau KM, Chan PH (1985) Aspects of the 40–50 day oscillation during the northern winter as inferred from outgoing longwave radiation. Mon Weather Rev 113(11):1889–1909

Lau WKM, Waliser DE (eds) (2012) Intraseasonal variability in the atmosphere–ocean climate system, 2nd edn. Springer, Berlin

Lin X, Johnson RH (1996) Kinematic and thermodynamic characteristics of the flow over the western Pacific warm pool during TOGA COARE. J Atmos Sci 53:695–715

Lin JL, Kiladis GN, Mapes BE, Weickmann KM, Sperber KR, Lin W, Wheeler M, Schubert SD, Del Genio A, Donner LJ, Emori S, Gueremy JF, Hourdin F, Rasch PJ, Roeckner E, Scinocca JF (2006) Tropical intraseasonal variability in 14 IPCC AR4 climate models Part I: Convective signals. J Climate 19:2665–2690

Madden RA, Julian PR (1971) Detection of a 40–50 day oscillation in the zonal wind in the tropical Pacific. J Atmos Sci 28(5):702–708

Madden RA, Julian PR (1972) Description of global-scale circulation cells in the Tropics with a 40–50 day period. J Atmos Sci 29:1109–1123

Madden RA, Julian PR (1994) Observations of the 40–50-day tropical oscillation—a review. Mon Weather Rev 122:814–837

Majda AJ (2003) Introduction to PDEs and waves for the atmosphere and ocean. Courant lecture notes in mathematics, vol 9. American Mathematical Society, Providence

Majda AJ, Biello JA (2004) A multiscale model for the intraseasonal oscillation. Proc Natl Acad Sci USA 101(14):4736–4741

Majda AJ, Stechmann SN (2009a) A simple dynamical model with features of convective momentum transport. J Atmos Sci 66:373–392

Majda AJ, Stechmann SN (2009b) The skeleton of tropical intraseasonal oscillations. Proc Natl Acad Sci USA 106(21):8417–8422

Majda AJ, Stechmann SN (2011) Nonlinear dynamics and regional variations in the MJO skeleton. J Atmos Sci 68:3053–3071

Majda AJ, Stechmann SN, Khouider B (2007) Madden–Julian Oscillation analog and intraseasonal variability in a multicloud model above the equator. Proc Natl Acad Sci USA 104(24):9919–9924

Malkus JS (1954) Some results of a trade-cumulus cloud investigation. J Meteorol 11(3):220–237

Maloney ED, Hartmann DL (1998) Frictional moisture convergence in a composite life cycle of the Madden–Julian oscillation. J Climate 11:2387–2403

Masunaga H, L'Ecuyer T, Kummerow C (2006) The Madden–Julian oscillation recorded in early observations from the tropical rainfall measuring mission (TRMM). J Atmos Sci 63(11):2777–2794

Matsuno T (1966) Quasi-geostrophic motions in the equatorial area. J Meteorol Soc Jpn 44(1):25–43

Myers D, Waliser D (2003) Three-dimensional water vapor and cloud variations associated with the Madden–Julian oscillation during Northern Hemisphere winter. J Climate 16(6):929–950

Nakazawa T (1988) Tropical super clusters within intraseasonal variations over the western Pacific. J Meteorol Soc Jpn 66(6):823–839

Roundy P, Frank W (2004) A climatology of waves in the equatorial region. J Atmos Sci 61(17):2105–2132

Salby M, Hendon H (1994) Intraseasonal behavior of clouds, temperature, and motion in the tropics. J Atmos Sci 51(15):2207–2224

Stechmann SN, Majda AJ (2015) Identifying the skeleton of the Madden–Julian oscillation in observational data. Mon Weather Rev 143:395–416. https://doi.org/10.1175/MWR-D-14-00169.1

Stechmann SN, Majda AJ, Skjorshammer D (2013) Convectively coupled wave–environment interactions. Theor Comput Fluid Dyn 27:513–532

Thual S, Majda AJ, Stechmann SN (2014) A stochastic skeleton model for the MJO. J Atmos Sci 71:697–715

Tian B, Waliser D, Fetzer E, Lambrigtsen B, Yung Y, Wang B (2006) Vertical moist thermodynamic structure and spatial–temporal evolution of the MJO in AIRS observations. J Atmos Sci 63(10):2462–2485

Tompkins AM (2001) Organization of tropical convection in low vertical wind shears: The role of water vapor. J Atmos Sci 58(6):529–545

Waite ML, Khouider B (2010) The deepening of tropical convection by congestus preconditioning. J Atmos Sci 67:2601–2615

Waliser D (2012) Predictability and forecasting. In: Lau WKM, Waliser DE (eds) Intraseasonal variability in the atmosphere–ocean climate system. Springer, Berlin

Wheeler M, Kiladis GN (1999) Convectively coupled equatorial waves: analysis of clouds and temperature in the wavenumber–frequency domain. J Atmos Sci 56(3):374–399

Wheeler M, Kiladis GN, Webster PJ (2000) Large-scale dynamical fields associated with convectively coupled equatorial waves. J Atmos Sci 57(5):613–640

Yanai M, Chen B, Tung WW (2000) The Madden–Julian oscillation observed during the TOGA COARE IOP: Global view. J Atmos Sci 57:2374–2396

Yang GY, Hoskins B, Slingo J (2007) Convectively coupled equatorial waves. Part I: Horizontal and vertical structures. J Atmos Sci 64(10):3406–3423

Yoneyama K, Zhang C, Long CN (2013) Tracking pulses of the Madden-Julian oscillation. Bull Am Meteor Soc 94(12):1871–1891

Zhang C (2005) Madden–Julian Oscillation. Rev Geophys 43:RG2003. https://doi.org/10.1029/2004RG000158

Zhang C, Hendon HH (1997) Propagating and standing components of the intraseasonal oscillation in tropical convection. J Atmos Sci 54(6):741–752

Chapter 3
A Stochastic Skeleton Model for the MJO

The Madden–Julian Oscillation (MJO) is actually a very irregular oscillation. Its oscillation period can vary significantly and is often described by a wide range such as 30–60 days or even 20–100 days. In addition, its amplitude is also irregular: sometimes the MJO is present, sometimes it is weak, and sometimes it is completely absent. Such irregularities present formidable challenges for modeling and prediction (and for planning expensive and intricate field campaigns to obtain in situ observations of an oscillation that may be temporarily in hibernation!). To account for these irregularities, a stochastic version of the skeleton model was developed by Thual et al. (2014) and is described in this chapter. The stochastic skeleton model is able to produce both the intermittent generation of MJO events and the organization of MJO events into wave trains of two or three successive events, as in observations. With such intrinsic irregular behavior in the model, it is possible that the stochastic skeleton model could be used for MJO forecasting. Other steps in the direction of MJO skeleton forecasting include the design of novel data assimilation algorithms for the skeleton model by Chen and Majda (2016) and the design of indices, based on MJO skeleton theory, for monitoring and predicting the MJO, which is described further in Chap. 5.

3.1 Introduction

As already discussed in the previous chapter, the Madden-Julian Oscillation (MJO, Madden and Julian 1971, 1994) in the tropics is an equatorial planetary-scale wave that also appears as a planetary envelope of several synoptic and mesoscale convective processes. An illustration of this feature is shown in Fig. 3.1. MJO events typically begin in the Indian Ocean (80–120°E) and propagate eastward in the western Pacific ocean (120°E–160°W) at around $5\,\mathrm{m\,s^{-1}}$ (Zhang 2005). In the passing of the MJO, many convective events develop that are of smaller size

© The Author(s), under exclusive licence to Springer Nature Switzerland AG 2019   29
A. J. Majda et al., *Tropical Intraseasonal Variability and the Stochastic Skeleton Method*, SpringerBriefs in Mathematics of Planet Earth, https://doi.org/10.1007/978-3-030-22247-5_3

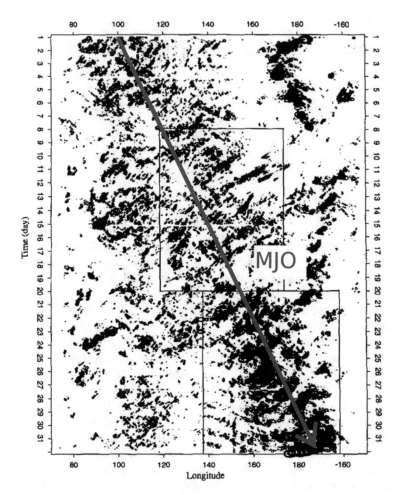

Fig. 3.1 A Nakazawa cloud-cluster diagram showing the propagation of an MJO event in nature and its convective features. The MJO propagation at around 5 m s^{-1} is indicated by the red arrow, as a function of longitude (degrees East) and time (days from 1 Dec 1992). The convective areas (defined here as Infrared Brightness Temperature <205 K) are shaded. From Chen et al. (1996); ©American Meteorological Society; used with permission

and lifespan, propagating either eastward or westward. Those events consist of a complex menagerie of convectively coupled equatorial waves, such as 2-day waves, convectively coupled Kelvin waves, etc. (Kiladis et al. 2009). They are favored and modulated by the planetary conditions associated to the MJO and contribute to it in return, resulting in a multiscale interaction.

An important feature of those synoptic and mesoscale convective processes is their apparent randomness and disorganization. In this chapter, we will account for this feature in the skeleton model for the MJO using a stochastic parameterization (Majda et al. 2008). Recall from the previous chapter that, in the skeleton model, the MJO results from a simple multiscale interaction between (1) planetary-scale dynamics, (2) moisture, and (3) the planetary envelope of synoptic convective activity. The rather complex details of synoptic convective activity, as shown in Fig. 3.1, are however unresolved. With respect to their planetary envelope, the contribution of those convective processes appears to be highly irregular, intermittent, and to have a low predictability. To account for this intermittent contribution while keeping the minimal design of the skeleton model (i.e., without solving entirely the synoptic details), one strategy is to develop a suitable stochastic parameterization. This stochastic parameterization will be introduced in the next sections.

With the incorporation of this stochastic parameterization, we will be able to simulate a more intermittent and irregular MJO as in nature. Recall the fundamental features of the MJO that are captured by the skeleton model from the previous chapter (Hendon and Salby 1994; Wheeler and Kiladis 1999; Zhang 2005):

I. A slow eastward phase speed of roughly $5 \, \text{m s}^{-1}$,
II. A peculiar dispersion relation with $d\omega/dk \approx 0$, and
III. A horizontal quadrupole structure.

While the features (I–III) are the salient intraseasonal-planetary features of MJO composites, individual MJO events are often unique, with, for example, varying structure, lifespan, intensity, and convective characteristics. In addition to the above features, the stochastic skeleton model introduced in this chapter also captures:

IV. The intermittent generation of MJO events, and
V. The organization of MJO events into wave trains with growth and demise.

A brief illustration of features (IV–V) is shown in Fig. 3.2. During this observation campaign there was an observed wave train of three consecutive MJO events (MJO-1 to 3) followed by a pause and an isolated event (MJO-4). This shows that the MJO in nature is not a perfectly regular cycle. Instead, the MJO events tend to organize into wave trains—i.e., into a series of successive MJO events (two, three, or more in a row). Understanding this organization remains an elusive scientific question sparking great interest. For example, primary MJO events at the beginning of a wave train have sometimes clearly identified precursors and sometimes few or none (Matthews 2008; Straub 2013). They may either be generated by the internal variability of certain tropical processes or as a secondary response to independently existing extratropical forcings (Zhang 2005; Lau and Waliser 2012).

The present chapter is organized as follows. The main features of the stochastic skeleton model are illustrated in Sect. 3.2. This includes an analysis of spacetime plots called Hovmollers (Sect. 3.2.1), MJO wave trains (Sect. 3.2.2), and power spectra (Sect. 3.2.3) in the model. Additional work that extends the stochastic skeleton model is also briefly discussed (Sect. 3.2.4). The derivation of the stochastic skeleton model is detailed in Sect. 3.3. Finally, Sect. 3.4 is a short concluding discussion to close the present chapter.

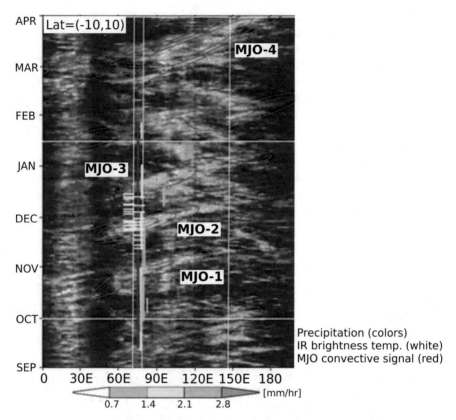

Fig. 3.2 Example of an observed MJO wave train during the CINDY-DYNAMO campaign of 2011. Hovmollers of several MJO indicators, as a function of longitude (deg East) and time (months), such as Precipitation (colors), Infrared Brightness Temperature (white), and an MJO convective index (red), all averaged from 10S to 10N. From Yoneyama et al. (2013); ©American Meteorological Society; used with permission

3.2 Features of the Stochastic Skeleton Model

3.2.1 Hovmoller (SpaceTime) Diagrams

We consider here numerical simulations with the stochastic skeleton model (Thual et al. 2014). The effect of the stochastic fluctuations is to create a realistic intermittency in the simulated MJO. Figure 3.3 shows the long-term evolution of zonal winds, potential temperature, moisture, and convective activity. The eastward propagation of several MJO events at around $5\,\mathrm{m\,s}^{-1}$ (red line) is visible on all fields, except potential temperature for which the MJO has a weak signature. In particular, the MJO appears here as an envelope of synoptic scale structures as seen by the smaller scale bursts along the tracks of propagation, which compares realistically to

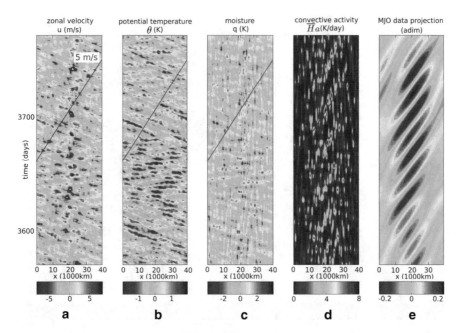

Fig. 3.3 Solutions of the stochastic skeleton model. Hovmoller diagrams of (**a**) zonal winds u (m s^{-1}), (**b**) potential temperature θ (K), (**c**) moisture q (K), and (**d**) convective activity $\overline{H}a$ (K day^{-1}) at the equator, as well as (**e**) an MJO data projection index (adim), as a function of zonal location x (1000 km) and time t (in days from an arbitrary reference time). Red lines indicate eastward propagation at 5 m s^{-1}

the behavior observed in nature (see Fig. 3.1). There are in addition some large-scale and small-scale propagating structures corresponding to other linear waves in the skeleton model discussed in the previous chapter (dry Kelvin, dry Rossby, or moist Rossby mode). In the right-hand side of Fig. 3.3, we show in addition an MJO index that evaluates the MJO intensity by projection of data on the linear solutions of the skeleton model (Majda and Stechmann 2011; Stechmann and Majda 2015). While in Stechmann and Majda (2015) observational data is projected, here we project the numerical solutions of the stochastic skeleton model that are in addition filtered in the intraseasonal-planetary band ($1 \leq k \leq 3$, $1/90 \leq \omega \leq 1/30$ cpd). This representation, along with comparison to the other Hovmollers diagrams shown in Fig. 3.3, allows us to identify clearly the MJO variability despite the noisy signals.

3.2.2   MJO Wave Trains

We focus now on the features of MJO wave trains in the stochastic skeleton model. Figure 3.4 repeats the Hovmollers over an extended period of time. The MJO events are organized into wave trains with growth and demise, i.e. into series of

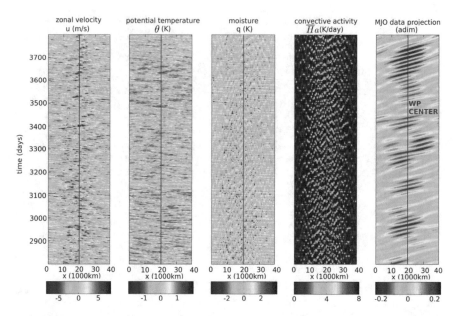

Fig. 3.4 Solutions of the stochastic skeleton model. As in previous Fig. 3.3 but for an extended period. Red lines indicate the center of the background warm pool state with increased cooling and moistening

successive MJO events following a primary MJO event. One series typically consists of a succession of either two, three, or four MJO events in a row. There is, for example, a series of two events around time 2900–3000 days, an isolated event around 3100 days, a series of three events around 3200–3400 days, and a series of four events around 3600–3900 days. This behavior compares realistically with the one found in observations in Fig. 3.2.

In order to show more examples of MJO wave trains, Fig. 3.5 repeats previous Hovmollers over an extended period of time showing only the MJO index. The MJO wave trains are intermittent with a great diversity in strength, structure, lifetime, and localization. There are in addition prolonged periods of moderate or weak activity between those wave trains. This is an attractive feature of the present stochastic skeleton model in generating MJO variability. This intermittency and organization results directly from the stochastic fluctuations: for instance, a useful comparison can be made with solutions of the deterministic skeleton model (Fig. 2.7 of Chap. 2) where the MJO variability is more regular. The MJO wave trains in Fig. 3.5 are overall centered on the region $x = 20,000$ km (red line): this is because, as in the previous chapter, this area represents the warm pool in nature and therefore has increased background cooling s_θ and moistening s_q. There are however some examples of wave trains stalling at the warm pool center or developing slightly outside the warm pool region, a situation that is also commonly found in nature (Wang and Rui 1990; Zhang and Hendon 1997). Note that the background warm

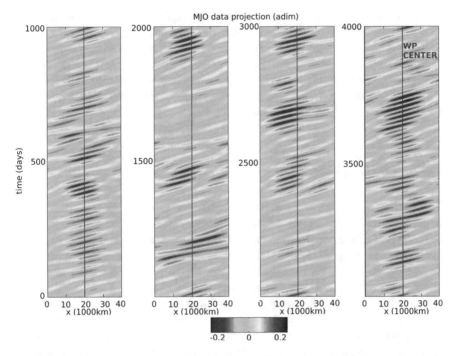

Fig. 3.5 Solutions of the stochastic skeleton model with examples of MJO wave trains. As in previous Figs. 3.3 and 3.4 but for an extended period and showing only the MJO data projection index. Red lines indicate the center of the background warm pool state with increased cooling and moistening

Table 3.1 Statistics of MJO wave trains in observations and the stochastic skeleton model

Event type	Observations (1979–2012)	Stochastic skeleton model (34 years)
Primary	154	98
Continuing	330	343
Terminal	154	98
Circumnavigating	15	24
Average duration of MJO events	39.7	34.8

Adapted from Stachnik et al. (2015)

pool region confines the MJO wave trains but is non-essential to their generation, as was shown in experiments with homogeneous background conditions (Thual et al. 2014).

In addition to the above features, the statistics of the MJO wave trains in the stochastic skeleton model compare very well with observations. Here we present Table 3.1 from Stachnik et al. (2015) that compares the occurrence of several types of MJO events in observations and a simulation of the present stochastic skeleton model. In particular, the occurrence of each type of events is found to be of a similar order of magnitude. Here by definition an MJO wave train consists of a first primary MJO event followed by a series of continuing events and finally a terminal event. Meanwhile, circumnavigating MJOs circle around the entire equator (Matthews 2008). In addition to this, the average duration of MJO events (of any given type) is similar. It is remarkable that such a simple stochastic model produces such good results while many general circulation models struggle to have any MJOs (see Stachnik et al. 2015 for a detailed discussion).

3.2.3   Power Spectra

Another realistic feature captured by the stochastic skeleton model is the distribution of energy across scales. Figure 3.6 shows the power spectra for the model's solutions as a function of the zonal wavenumber k (in $2\pi/40{,}000\,\text{km}$) and frequency ω (in cycle per days or cpd, 1/30 cpd, for example, corresponds to a period of 30 days). The power spectra is obtained by decomposing the model's solutions in Fourier space in both space and time and computing energy for each k and ω. The zonal wavenumber k in particular corresponds to lengths that are a fraction of the equator's periodic circumference ($k = 1$ corresponds to a sinusoid of length 40,000 km circling the equator, $k = 2$ to 20,000 km, etc.). Such a representation has gained widespread popularity in the atmospheric community because the distribution of power for several convective signals found in nature compares very well with equatorial waves theory (Wheeler and Kiladis 1999; Kiladis et al. 2009). Note however that most models, no matter their level of complexity, struggle to reproduce a realistic power spectra for the MJO (Lin et al. 2006).

   The power spectra of the skeleton model in contrast agree very well with nature: for instance, a useful comparison can be made between Fig. 3.6 and the observed power spectra of Outgoing Longwave Radiation (an analogue of convective activity, Wheeler and Kiladis 1999; Stechmann and Majda 2015). However, such a comparison with observations is only valid at interannual and planetary scales ($|k| \leq 5$ and $|\omega| \leq 1/30$ cpd) due to the starting assumptions of the skeleton model that omit (i.e., parameterize) synoptic and mesoscale details. A realistic feature in Fig. 3.6 is the maximal power in the MJO band, which is usually defined as the band $1 \leq k \leq 3$ and $1/90 \leq \omega \leq 1/30$ cpd. This power peak is seen on all variables except potential temperature upon which the MJO has a weak signature. Such a power peak roughly corresponds to the slow eastward phase speed of $\omega/k \approx 5\,\text{m s}^{-1}$ with the peculiar dispersion relation $d\omega/dk \approx 0$ described at the beginning of this chapter. In addition, the eastward propagation of the MJO is not compensated by westward propagations, as can be seen by weaker power in the antisymmetric band $-3 \leq k \leq -1$ and $1/90 \leq \omega \leq 1/30$ cpd. This distribution of power is related to some extent to the linear solutions of the deterministic skeleton model (see Chap. 2): for instance, power in Fig. 3.6 tends to be maximal near the dispersion curves of those linear solutions (black dots) that consist of the MJO mode, moist Rossby mode as well as dry Kelvin and Rossby modes at higher frequencies outside of the range

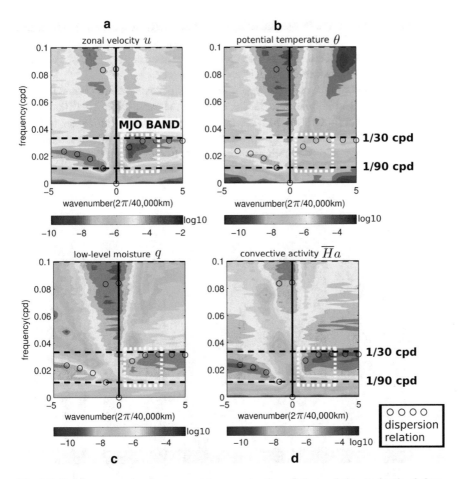

Fig. 3.6 Zonal wavenumber-frequency power spectra for solutions of the stochastic skeleton model. For (**a**) zonal winds u (m s^{-1}), (**b**) potential temperature θ (K), (**c**) moisture q (K), and (**d**) convective activity $\overline{H}a$ (K day^{-1}) taken at the equator, as a function of zonal wavenumber (in $2\pi/40,000$ km) and frequency (in cycles per day or cpd). The contour levels are in the base-10 logarithm, for the dimensional variables taken at the equator. The black circles mark the dispersion curves from the model's linear solutions. The black dashed lines mark the periods 90 and 30 days. The white dashed lines mark the MJO band ($k = 1 - 3$, $\omega = 1/90 - 1/30$ cpd)

of the graph. Note finally that the zonal shape of the background warm pool region slightly favors power at wavenumber $k = 1$, however a similar power spectra is found even in experiments with homogeneous background conditions (not shown, see Thual et al. 2014).

3.2.4 Beyond the Stochastic Skeleton Model

While we have illustrated here the main features of the stochastic skeleton model, a large body of work extends beyond this simple scope. Because the skeleton model consistently captures the most salient features of the MJO, it provides an invaluable theoretical framework to address more complex problems of intraseasonal variability in general. Here a few examples of those applications are briefly discussed that include the stochastic skeleton model parameterization.

Figure 3.7 shows the long-term evolution of a stochastic skeleton model where we have included in simple fashion the effect of the seasonal cycle (Thual et al. 2015). In this experiment, the background warm pool state (with center in black line) is displaced meridionally following the seasons (to the northern hemisphere in boreal summer and southern hemisphere in boreal winter), which mimics

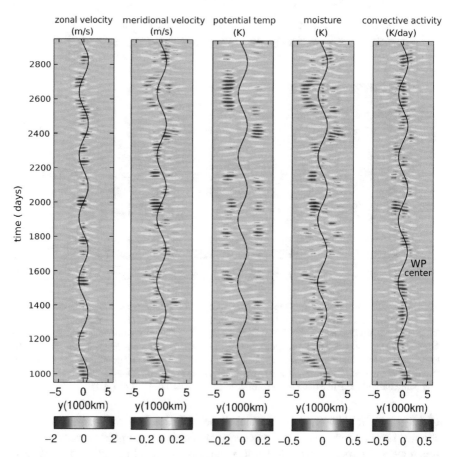

Fig. 3.7 Stochastic Skeleton Model with Seasonal Cycle. Hovmoller diagrams: for (**a**) zonal winds u (m s^{-1}), (**b**) meridional winds v (m s^{-1}), (**c**) potential temperature θ (K), (**d**) moisture q (K), and (**e**) convective activity $\overline{H}a$ (K day^{-1}), as a function of meridional position location y (in 1000 km) and simulation time (in 1000 days). The variables are filtered in the MJO band ($1 \leq k \leq 3$ and $1/90 \leq \omega \leq 1/30$ cpd), and considered at the warm pool zonal center ($x = 20{,}000$ km). The meridional position of the warm pool center, varying with seasons, is overplotted (black line)

qualitatively the observed migration of warm sea surface temperatures (Zhang and Dong 2004). This march of seasons has important implications for the MJO in nature that is most prominent in boreal winter but of a different nature in boreal summer, which the model intends to capture. In particular, there are additional types of intraseasonal wave trains in Fig. 3.7 beyond the MJO wave trains that can develop off-equator with unique meridional structures. Some of those wave trains propagating northward in boreal summer (e.g., at time 2800 days) are potential surrogates for intraseasonal events initiating the monsoon over the Asian continent in nature (Lawrence and Webster 2002). Note that to compute the solutions in Fig. 3.7 the skeleton model has been extended to include more meridional modes in order to allow for off-equatorial convective activity (see Thual et al. 2015 for details).

Many more examples of applications beyond the skeleton model and its stochastic version are detailed in the next chapters. In Chap. 5, for example, a realistic background warm pool state is included in the model (Ogrosky and Stechmann 2015). In Chap. 6 the skeleton model is extended to reproduce the complex vertical tilted structure of the MJO in nature (Thual and Majda 2015). Importantly, the realistic features discussed above such as the MJO power spectra and its organization into intermittent wave trains are still consistently retrieved in those works despite the increased level of complexity and different scope. This illustrates the high robustness of the skeleton model's theoretical framework in practical settings.

3.3 Derivation of the Stochastic Skeleton Model

3.3.1 Markov Birth-Death Process

In this section we detail the derivation of the stochastic skeleton model. In the skeleton model, the MJO results from a simple multiscale interaction between (1) the planetary-scale dynamics, (2) moisture and (3) the planetary envelope of synoptic activity (see discussion above). With respect to the planetary-scale processes depicted in the skeleton model, the contribution of those synoptic details appears to be highly irregular, intermittent, and have a low predictability. To account for this intermittent contribution while keeping the minimal design of the skeleton model (i.e., without solving entirely the synoptic details), we develop here a suitable stochastic parameterization that consists of a Markov birth-death process.

We first briefly present some theoretical aspects of Markov birth-death processes. Those are the simplest type of continuous-time Markov processes (see, e.g., chapter 7 of Gardiner 1994; Lawler 2006 for an introduction). Let the envelope of synoptic activity a be a random variable taking discrete values $a = \Delta a\, \eta$ with η an integer. The probabilities of transiting from one state η to another over a time step Δt read as follows:

Fig. 3.8 Example of evolution of a Markov birth-death process for the variable $a(t)$. The variable a (red) transits intermittently between discrete levels of interval Δa, according to the upward and downward transition rates λ and μ, respectively

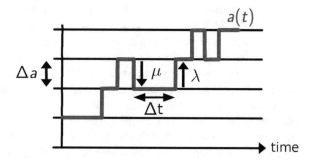

$$
\begin{aligned}
P\{\eta(t + \Delta t) = \eta(t) + 1\} &= \lambda \Delta t + o(\Delta t) \\
P\{\eta(t + \Delta t) = \eta(t) - 1\} &= \mu \Delta t + o(\Delta t) \\
P\{\eta(t + \Delta t) = \eta(t)\} &= 1 - (\lambda + \mu)\Delta t + o(\Delta t) \\
P\{\eta(t + \Delta t) \neq \eta(t) - 1, \ \eta(t), \ \eta(t) + 1\} &= o(\Delta t),
\end{aligned}
\tag{3.1}
$$

where λ and μ are the upward and downward rates of transition, respectively. An illustration of the evolution of a is shown in Fig. 3.8. The birth-death process allows for intermittent transitions between the states η: in particular, a can intermittently increase at rate λ or decrease at rate μ. The transition rates λ and μ have to be specified, as discussed hereafter. In order to obtain non-trivial solutions those transition rates usually depend on the state of the system (e.g., on η or additional variables when coupled to other processes).

In addition to the above definition, Markov birth-death processes have interesting tractable properties. For example, such a process can alternatively be expressed in the form of a master equation:

$$
\partial_t P(\eta) = [\lambda(\eta-1)P(\eta-1) - \lambda(\eta)P(\eta)] + [\mu(\eta+1)P(\eta+1) - \mu(\eta)P(\eta)], \tag{3.2}
$$

where $P(\eta)$ is the probability of the state η (not to be mistaken with the conditional probabilities in Eq. (3.1)). While the above equation can be difficult to solve in practice due to its high dimensionality, useful insight can be found by computing, for example, the evolution of the first moment (i.e., the mean-field equation). This reads simply:

$$
\partial_t E(\eta) = E(\lambda - \mu), \tag{3.3}
$$

where the $E(\eta)$ denote statistical expected value (i.e., mean). In particular, in the asymptotic limit of small transitions (Δa small) the above equation is deterministic, due to $E(x) \approx x$.

3.3.2 Choice of Transition Rates

Next, we need to choose suitable transition rates λ and μ that are consistent with the dynamics of the skeleton model. Here, the design principle is that such dynamics must be recovered on average. One simple and practical choice of the transition rates that satisfies this design principle reads:

$$\lambda = \begin{cases} \Gamma |q|\eta + \delta_{\eta 0} \text{ if } q \geq 0 \\ \delta_{\eta 0} \text{ if } q < 0 \end{cases} \text{ and } \mu = \begin{cases} 0 \text{ if } q \geq 0 \\ \Gamma |q|\eta \text{ if } q < 0. \end{cases} \tag{3.4}$$

The justification for this choice is as follows. First, we need to ensure $a \geq 0$ as in the original skeleton model with $a = \Delta a \eta$. For this, the Kronecker delta operator $\delta_{\eta 0}$ ($\delta_{\eta 0} = 1$ if $\eta = 0$ and $\delta_{\eta 0} = 0$ otherwise) ensures that $\lambda \geq 1$ at the boundary condition $\eta = 0$ such that $\eta \geq 0$ remains a positive integer. Second, the associated mean-field equation reads:

$$\partial_t E(a) = \Delta a E(\lambda + \mu) = \Gamma E(qa) + \Delta a E(\delta_{\eta 0}), \tag{3.5}$$

which, in the asymptotic limit of small transitions Δa (for which $E(x) \approx x$) is identical to the deterministic evolution of convective activity a from previous Chap. 2:

$$\partial_t a = \Gamma qa. \tag{3.6}$$

As a result, we conserve the essential properties of the skeleton model with this stochastic parameterization.

3.3.3 Gillespie Algorithm

In order to compute numerical solutions of the above Markov birth–death process we use a Gillespie algorithm (Gillespie 1975, 1977), which is detailed here. Instead of solving the entire master equation (3.2) for the probabilities of each state, we solve in practice only one realization of the system according to the sequential update:

$$a(t + \tau) = a(t) + \xi \Delta a. \tag{3.7}$$

where τ and ξ are random variables with the following cumulative distribution functions:

$$P(\tau) = \exp(-(\lambda + \mu)\tau)$$
$$P(\xi) = \{\mu/(\lambda + \mu), \lambda/(\lambda + \mu)\}, \tag{3.8}$$

that depend on the transition rates λ and μ given earlier. Given $a(t)$ at a time t, we first try for the random variable $\tau \geq 0$ that is the time interval before the next consecutive transition (with distribution following a Poisson law). Next, we try for the random variable $\xi = \{-1, 1\}$ that gives the direction of transition (either upward or downward). This procedure provides the conditions at the next step $a(t + \tau)$ and is then repeated sequentially. If we compare this process to a purely deterministic evolution (e.g., as in Eq. (3.6)) we see that the Markov birth-death process introduces randomly some additional advances and delays through τ as well as some occasional opposite transitions through ξ. If $\lambda > \mu$, for example, there will be an increased likelihood of an upward transition, though a downward transition is still possible.

3.3.4 Stochastic Single-Column Skeleton Model

We analyze here single-column versions of the skeleton model. Such a simplification provides useful insight on the basic oscillation mechanism in the skeleton model as well as its modification by the Markov birth-death process. In particular, it is shown that the generation of MJO wave trains is consistent with noise-induced chaos, which has been shown in other settings to govern the overall evolution of the stochastic skeleton model (Majda and Tong 2015). First, let $\partial_x = \partial_y = 0$ in the deterministic skeleton model from previous chapter 1 such that we omit the role of propagating solutions and focus on the dynamics at a single location. The resulting deterministic single-column model reads:

$$\partial_t q = -\overline{H}a + s^q$$
$$\partial_t a = \Gamma q a. \tag{3.9}$$

In contrast, the stochastic single-column model where the above Markov birth-death process governs the evolution of convective activity reads:

$$\partial_t q = -\overline{H}a + s^q$$
$$a(t + \tau) = a(t) + \xi \Delta a. \tag{3.10}$$

A comparison of solutions for both systems is shown in Fig. 3.9. As shown in the time series, the deterministic single-column model simulates a regular oscillation with frequency $\omega = \sqrt{\Gamma s^q}$ (black). Such an oscillation is slightly asymmetric due to the simple nonlinearity in Eq. (3.9), resulting in sharp and intense maximas of a. In the complete skeleton model, this basic oscillation between moisture and convective activity coupled to planetary dynamics gives the overall structure and propagation of the MJO. Interestingly, this oscillation conserves a Hamiltonian $H = \Gamma q^2/2 + \overline{H}a - \ln(a)s^q$ that satisfies $\partial_t q = -\partial_l H$ and $\partial_t l = \partial_q H$ where $l = \log(a)$, along with $\partial_t H = 0$. As a result, solutions of the deterministic single-column model in Fig. 3.9 remain on a single iso-Hamiltonian level in the phase space $q - a$ (black), with value given by the initial conditions.

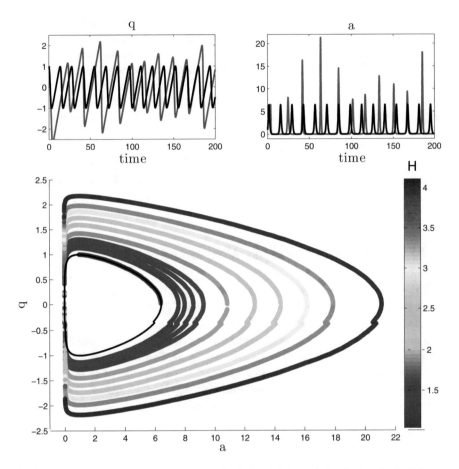

Fig. 3.9 Comparison of solutions between the Deterministic and Stochastic Single-Column Skeleton Model. (Top): Time series of q (top left) and a (top right), as a function of time t for the deterministic (black) and stochastic version (red). (Bottom): Scatterplot of Hamiltonian H in phase space $a - q$, for the deterministic (black) and stochastic version (colors). All variables are nondimensional

Focus now on the time series of the stochastic single-column model in Fig. 3.9. In contrast, this system shows an irregular and intermittent oscillation with a modulated amplitude (red), that approximately conserves the frequency $\omega = \sqrt{\Gamma s^q}$. In the complete stochastic skeleton model, this irregular oscillation coupled to planetary-scale dynamics is responsible to some extent for the generation of intermittent MJO wave trains with growth and demise, as discussed in previous sections. As shown in Fig. 3.9 in the phase space $q - a$ (colors), this irregular oscillation does not conserve a Hamiltonian but rather transits intermittently between different iso-Hamiltonian levels which explains its modulated amplitude. Those transitions are due to small perturbations introduced by the Markov birth-death process that violate the Hamiltonian conservation. For example, the random variable τ may introduce some random advances or delay as compared to a purely deterministic evolution, while the random variable ξ may introduce some occasional opposite transitions. For instance, recall from Eq. (3.5) that the Markov birth-death process conserves the deterministic dynamics only on average. Those perturbations do not need to be large to disrupt the oscillation in an effective fashion: for instance, as shown in Fig. 3.9 when a becomes small the iso-Hamiltonians are very close, therefore small perturbations of a greatly modify the Hamiltonian and as a result the amplitude of the next oscillation. Such a behavior with a high sensitivity of the large-scale evolution to stochastic fluctuations is consistent with noise-induced chaos, which has been shown in other settings to govern the overall evolution of the stochastic skeleton model (Majda and Tong 2015).

3.3.5   Complete Stochastic Skeleton Model

We show here the complete formulation of the stochastic skeleton model. Recall from previous chapters that, in the skeleton model, the MJO results from a simple multiscale interaction between (1) planetary-scale dynamics, (2) moisture, and (3) the planetary envelope of synoptic convective activity. In the stochastic skeleton model, the evolution of (3) convective activity is driven by the Markov birth-death process described above. As a result, the contribution of those convective processes to their planetary envelope is highly irregular, intermittent, and has a low predictability. The stochastic skeleton model reads:

$$
\begin{aligned}
&\partial_t u - yv - \partial_x \theta = 0 \\
&yu - \partial_y \theta = 0 \\
&\partial_t \theta - (\partial_x u + \partial_y v) = \overline{H}a - s^\theta \\
&\partial_t q + \overline{Q}(\partial_x u + \partial_y v) = -\overline{H}a + s^q \\
&a(t + \tau) = a(t) + \xi \Delta a,
\end{aligned}
\tag{3.11}
$$

with zonal and meridional velocity u and v, respectively, potential temperature θ, low-level moisture q, and convective activity a. All variables are anomalies from a radiative-convective equilibrium, except a. The last row in particular describes the Gillespie algorithm from the previous subsection used to solve one realization of the Markov birth-death process, with random variables τ and ξ that depend on the system's state. This model contains a minimal number of parameters: $\overline{Q}$ is the background vertical moisture gradient, Γ is a proportionality constant, and $\overline{H}$ defines a heating/drying rate $\overline{H}a$ for the system in dimensional units. The s^θ and s^q are external sources of cooling and moistening, respectively, that need to be prescribed in the system. The solutions of such a system are presented in previous sections (see, e.g., Figs. 3.3, 3.4 and 3.5).

Generally speaking, the formulation of the above stochastic skeleton model follows a prototype found in several previous studies (Majda et al. 2008). The methodology consists in coupling some simple stochastic triggers (e.g., birth/death, spin-flip, coarse-grained lattice models...) to the otherwise deterministic processes, according to some probability laws motivated by physical intuition gained (elsewhere) from observations and detailed numerical simulations (Gardiner 1994; Katsoulakis et al. 2003; Lawler 2006). The methodology has been successful in parameterizing with more realism some essential processes of tropical variability for which high irregularity, high intermittency, and/or low predictability is involved (Majda et al. 2008).

The numerical method used to solve the stochastic skeleton model is as follows (Thual et al. 2014). In practice the system from Eq. (3.11) is truncated meridionally to the first parabolic cylinder function $\phi_0(y)$ (see Chap. 2), therefore we solve for the evolution of the variables $K(x, t)$, $R(x, t)$ as well as $a(x, t)$ and $Z(x, t) = q + \overline{Q}\theta$ considered in a zonal strip along the equator ($y = 0$). We have in particular $q = Q\phi_0(0)$, $a = A\phi_0(0)$, $s^\theta = S^\theta\phi_0(0)$, etc. as well as $\theta = -[K + R]\phi_0(0)$. The truncated system with those variable changes reads:

$$
\begin{aligned}
&\partial_t K + \partial_x K = (S^\theta - \overline{H}A)/2 \\
&\partial_t R - \partial_x R/3 = (S^\theta - \overline{H}A)/3 \\
&\partial_t Z = (1 - \overline{Q})(s^q - \overline{H}a) \\
&a(t + \tau) = a(t) + \xi \Delta a.
\end{aligned}
\tag{3.12}
$$

(Note that a slightly different scaling of the equations is used here in comparison with Chap. 2.) The spatial and temporal resolution consists of a spatial step Δx of 625 km spanning the equatorial belt (40,000 km) and a timestep ΔT of around 1.7 h. We use a splitting method to update the system over each timestep ΔT. First, Z and a in the zonal strip are held fixed and we solve for the evolution of K and R exactly using zonal Fourier series. Second, K and R are held fixed and we solve for the evolution of Z and a together as a series of consecutive transitions over smaller timesteps τ. The last consecutive transition in particular usually occurs after the end of the timestep ΔT, and is therefore omitted as an approximation in order to retrieve $Z(t + \Delta T)$ and $a(t + \Delta T)$.

The reference parameters values used to compute figures in this chapter read, in nondimensional units: $\overline{Q} = 0.9$, $\Gamma = 1.66$ (≈ 0.3 K^{-1} day^{-1}), $\overline{H} = 0.22$ (10 K day^{-1}), with stochastic transition parameter $\Delta a = 0.001$. The background warm pool state of cooling/moistening is defined by the external sources $s^\theta = s^q = 0.022(1 - 0.6\cos(2\pi x/L))$ at the equator, where $L = 40,000$ km is the equatorial belt length. For Fig. 3.9 we use instead $s^q = 0.22$ in order to mimic an MJO oscillation period around 40 days despite the single-column simplification. The dimensional reference scales are x, y: 1500 km, t: 8 h, u: 50 m s^{-1}, θ, q: 15 K (see Table 1 of Stechmann et al. 2008).

3.4   Concluding Discussion

To conclude, we have shown in this chapter how a suitable stochastic parameterization of the skeleton model allows for capturing additional realistic features of the MJO in nature. The skeleton model presented in Chap. 2 already captures realistically together the MJO's salient features of (I) a slow eastward phase speed of roughly 5 m s^{-1}, (II) a peculiar dispersion relation with $d\omega/dk \approx 0$, and (III) a horizontal quadrupole structure (Majda and Stechmann 2009, 2011). In addition to those features, the inclusion of a stochastic parameterization for convective processes allows the model to capture:

IV. The intermittent generation of MJO events, and
 V. The organization of MJO events into wave trains with growth and demise.

The solutions of such a stochastic skeleton model are in agreement with nature in many ways. In particular, the MJO events in the model are organized into intermittent wave trains (i.e., series) with a great diversity in strength, structure, lifetime, and localization. In addition to this, such a model succeeds at the difficult test of recovering a realistic power spectrum with a consistent distribution of energy across spatial scales and time scales for the MJO. We have also provided in this chapter a detailed derivation explaining the underlying mathematical aspects of such a model and their practical implementation. Because of its realism, the skeleton model also provides an invaluable theoretical framework to address more complex problems of intraseasonal variability in general. While we have illustrated here the main features of the stochastic skeleton model, a large body of exciting work will be shown in the next chapters that extend beyond this simple scope.

References

Chen N, Majda AJ (2016) Filtering the stochastic skeleton model for the Madden–Julian oscillation. Mon Weather Rev 144(2):501–527

Chen SS, Houze RA Jr, Mapes BE (1996) Multiscale variability of deep convection in relation to large-scale circulation in TOGA COARE. J Atmos Sci 53(10):1380–1409

Gardiner CW (1994) Handbook of stochastic methods for physics, chemistry, and the natural sciences. Springer, Berlin, 442 pp

Gillespie DT (1975) An exact method for numerically simulating the stochastic coalescence process in a cloud. J Atmos Sci 32:1977–1989. https://doi.org/10.1175/1520-0469(1975)032<1977:AEMFNS>2.0.CO;2

Gillespie DT (1977) Exact stochastic simulation of coupled chemical reactions. J Phys Chem 81(25):2340–2361

Hendon HH, Salby ML (1994) The life cycle of the Madden-Julian oscillation. J Atmos Sci 51:2225–2237

Katsoulakis MA, Majda AJ, Vlachos DG (2003) Coarse-grained stochastic processes for microscopic lattice systems. Proc Natl Acad Sci USA 100(3):782–787. https://doi.org/10.1073/pnas.242741499

Kiladis GN, Wheeler C, Haertel PT, Straub KH, Roundy PE (2009) Convectively coupled equatorial waves. Rev Geophys 47:rG2003. https://doi.org/10.1029/2008RG000266

Lau WM, Waliser DE (2012) Intraseasonal variability in the atmosphere-ocean climate system. Springer, Berlin, 642 pp

Lawler GF (2006) Introduction to stochastic processes. Chapman and Hall/CRC, New York, 192 pp

Lawrence DM, Webster PJ (2002) The boreal summer intraseasonal oscillation: relationship between northward and eastward movement of convection. J Atmo Sci 59:1593–1606

Lin JL, Kiladis GN, Mapes BE, Weickmann KM, Sperber KR, Lin W, Wheeler MC, Schubert SD, Del Genio A, Donner LJ, Emori S, Gueremy JF, Hourdin F, Rasch PJ, Roeckner E, Scinocca JF (2006) Tropical intraseasonal variability in 14 IPCC AR4 climate models. Part I: convective signals. J Climate 19:2665–2690. https://doi.org/10.1175/JCLI3735.1. http://journals.ametsoc.org/doi/abs/10.1175/JCLI3735.1

Madden RE, Julian PR (1971) Detection of a 40–50 day oscillation in the zonal wind in the tropical Pacific. J Atmos Sci 28:702–708

Madden RE, Julian PR (1994) Observations of the 40–50 day tropical oscillation-A review. Mon Weather Rev 122:814–837

Majda AJ, Stechmann SN (2009) The skeleton of tropical intraseasonal oscillations. Proc Natl Acad Sci 106:8417–8422. https://doi.org/10.1073/pnas.0903367106. http://www.pnas.org/content/106/21/8417

Majda AJ, Stechmann SN (2011) Nonlinear dynamics and regional variations in the MJO skeleton. J Atmos Sci 68:3053–3071. https://doi.org/10.1175/JAS-D-11-053.1. http://journals.ametsoc.org/doi/abs/10.1175/JAS-D-11-053.1

Majda A, Tong X (2015) Geometric ergodicity for piecewise contracting processes with application for tropical stochastic lattice models. Commun Pure Appl Math. https://doi.org/10.1002/cpa.21584

Majda AJ, Franzke C, Khouider B (2008) An applied mathematics perspective on stochastic modelling for climate. Philos Trans R Soc A: Math Phys Eng Sci 366(1875):2427–2453. https://doi.org/10.1098/rsta.2008.0012. http://rsta.royalsocietypublishing.org/content/366/1875/2427

Matthews AJ (2008) Primary and successive events in the Madden-Julian oscillation. Q J R Meteorol Soc 134:439–453

Ogrosky HR, Stechmann SN (2015) The MJO skeleton model with observation-based background state and forcing. Q J R Meterol Soc. https://doi.org/10.1002/qj.2552

Stachnik J, Waliser D, Majda A (2015) Precursor environmental conditions associated with the termination of Madden-Julian oscillation events. J Atmos Sci. http://dx.doi.org/10.1175/JAS-D-14-0254.1

Stechmann SN, Majda AJ (2015) Identifying the skeleton of the Madden-Julian oscillation in observational data. Mon Weather Rev 143:395–416

Stechmann SN, Majda AJ, Boulaem K (2008) Nonlinear dynamics of hydrostatic internal gravity waves. Theor Comput Fluid Dyn 22(6):407–432

Straub K (2013) MJO Initiation in the real-time multivariate MJO index. J Climate 26:1130–1151

Thual S, Majda AJ (2015) A skeleton model for the MJO with refined vertical structure. Clim Dyn 46. https://doi.org/10.1007/s00382-015-2731-x

Thual S, Majda AJ, Stechmann SN (2014) A stochastic skeleton model for the MJO. J Atmos Sci 71:697–715

Thual S, Majda AJ, Stechmann SN (2015) Asymmetric intraseasonal events in the skeleton MJO model with seasonal cycle. Clim Dyn. https://doi.org/10.1007/s00382-014-2256-8

Wang B, Rui H (1990) Synoptic climatology of transient tropical intraseasonal convection anomalies: 1975–1985*. Meteorol Atmos Phys 44:43–61

Wheeler M, Kiladis GN (1999) Convectively coupled equatorial waves: analysis of clouds and temperature in the wavenumber-frequency domain. J Atmos Sci 56:374–399. https://doi.org/10.1175/1520-0469(1999)056<0374:CCEWAO>2.0.CO;2

Yoneyama K, Zhang C, Long CN (2013) Tracking pulses of the Madden-Julian oscillation. Bull Am Meteor Soc. http://dx.doi.org/10.1175/BAMS-D-12-00157.1

Zhang C (2005) Madden-Julian oscillation. Rev Geophys 43:rG2003. https://doi.org/10.1029/
    2004RG000158
Zhang C, Dong M (2004) Seasonality in the Madden-Julian oscillation. J Clim 17:3169–3180
Zhang C, Hendon HH (1997) Propagating and standing components of the intraseasonal oscillation
    in tropical convection. J Atmos Sci 54:741–752

Chapter 4
Tropical–Extratropical Interactions and the MJO Skeleton Model

The MJO is a global phenomenon. Its strongest signature is in the tropics, but its structure includes Rossby gyres that extend significantly away from the equator. Correlations have been identified between the MJO and countless weather and climate phenomena around the globe. For instance, MJO activity is related with tropical phenomena such as monsoons, hurricanes, and El Niño–Southern Oscillation (ENSO), as well as extratropical phenomena such as tornadoes in the United States, Arctic sea ice, and heavy rain and snow events in the western United States, to name only a few. As a result, better forecasts of the MJO could have an impact on forecasts in the extratropics in, for example, the United States and Europe. Moreover, since the MJO is an intraseasonal oscillation with a period of 30–60 days, it plays a major role in the new frontier of subseasonal to seasonal prediction with lead times of weeks or months in advance.

What are the mechanisms behind the interactions between the MJO and the extratropics? Which other wave types and weather/climate phenomena are involved in the interactions? These questions are investigated in the present chapter in the nonlinear coupling between the MJO skeleton model (Chap. 2) and barotropic Rossby waves which have significant extratropical signature. A multiscale asymptotic analysis is used to derive resonant interaction equations for interactions between the MJO, Rossby waves (equatorial and extratropical), and the Walker circulation. In the multiscale analysis, the MJO and Rossby wave phase oscillations are the fast scales, and the different waves exchange energy on longer, slow time scales. In addition to studying the influence of the MJO on the midlatitudes or extratropics, this setup also allows investigation of the opposite direction: the influence of extratropical Rossby waves on the MJO. In particular, such extratropical influence is another important route to MJO initiation or termination, in addition to the tropical route that was investigated earlier via stochastic mechanisms (Chap. 3).

A. J. Majda et al., *Tropical Intraseasonal Variability and the Stochastic Skeleton Method*, SpringerBriefs in Mathematics of Planet Earth, https://doi.org/10.1007/978-3-030-22247-5_4

4.1 Introduction

Besides its strong tropical signal, the MJO interacts with the global flow on the intraseasonsal timescales. Teleconnection patterns between the global extratropics and the MJO have been described in early observational analyses by Weickmann (1983), Weickmann et al. (1985), and Liebmann and Hartmann (1984). Their results demonstrate coherent fluctuations between extratropical flow and eastward-propagating outgoing longwave radiation (OLR) anomalies in the tropics. In a later study, Matthews and Kiladis (1999) illustrate the interplay between high-frequency transient extratropical waves and the MJO. More recently, Weickmann and Berry (2009) demonstrate that convection in the MJO frequently evolves together with a portion of the activity in a global wind oscillation. Gloeckler and Roundy (2013) argued by using lagged composite analysis that the high amplitude extratropical circulation pattern is associated with simultaneous occurrence of both the MJO and the equatorial Rossby wave events.

Besides observational analyses, models have also been used to study the interactions between the MJO and extratropical waves. By including tropical convection forcing data in a barotropic model, Ferranti et al. (1990) found significant improvement in the model's predictability. Hoskins and Ambrizzi (1993) argued from their model that a zonally varying basic state is necessary for the MJO to excite extratropical waves by forcing perturbations to a barotropic model. To view the extratropical response to convective heating, Jin and Hoskins (1995) forced a primitive equation model with a fixed heat source in the tropics in the presence of a climatological background flow and obtain the Rossby wave train response as a result. To diagnose the specific response to patterns of convection like those of the observed MJO, Matthews et al. (2004) forced a primitive equation model in a climatological background flow with patterns of observed MJO. The resulting global response to that heating is similar in many respects to the observational analysis. The MJO initiation in response to extratropical waves was illustrated by Ray and Zhang (2010). They show that a dry-channel model of the tropical atmosphere developed MJO-like signals in tropical wind fields when forced by reanalysis fields at poleward boundaries. In addition, Lin et al. (2009) showed the significance of midlatitude dynamics in triggering tropical intraseasonal response by including extratropical disturbances in a tropical circulation model. Frederiksen and Frederiksen (1993) used a two-level primitive equation eigenvalue model and found large-scale basic-state flow and cumulus heating to be necessary for generating MJO modes with realistic structures. Many other interesting studies on tropical-extratropical interactions have been carried out. For example, see the review by Roundy (2011).

In past studies based on climate models, typically the effect of the MJO has been represented by forced perturbations (Hoskins and Ambrizzi 1993; Jin and Hoskins 1995; Matthews et al. 2004), or the influence of the midlatitude variations have been treated as boundary effects for the tropical circulation model (Frederiksen and Frederiksen 1993; Lin et al. 2009; Ray and Zhang 2010; Roundy 2011). Such simplifications are useful for isolating individual processes within these complex models.

As a next step, it would be desirable to design a simplified model where both the MJO and extratropical waves are simultaneously interactive, rather than externally imposing one of these two components; such an approach was recently taken by Chen et al. (2015, 2016), and the present chapter describes the results.

4.2 Three-Wave Interactions

To obtain an interactive tropical–extratropical model, while simultaneously including interactive effects of water vapor and convection, the MJO skeleton model of Chap. 2 is coupled nonlinearly to a model for barotropic Rossby waves which have significant extratropical signature. To analyze the nonlinear interactions, a multiscale asymptotic analysis is carried out to identify resonant wave interactions. The wave interactions can then be described in simplified form as a system of nonlinear ordinary differential equations (ODEs), which can be analyzed to understand the basic aspects of tropical–extratropical interactions of the MJO with interactive moisture.

4.2.1 The Two-Layer Equatorial Beta-Plane Equations with the MJO Skeleton

The nondimensional two-layer equatorial beta-plane equations for the barotropic and baroclinic MJO skeleton model are given by

$$\frac{\partial \bar{\mathbf{v}}}{\partial t} + \bar{\mathbf{v}} \cdot \nabla \bar{\mathbf{v}} + y\bar{\mathbf{v}}^\perp + \nabla \bar{p} = -\frac{1}{2} \nabla \cdot (\mathbf{v} \otimes \mathbf{v}), \tag{4.1a}$$

$$\nabla \cdot \bar{\mathbf{v}} = 0, \tag{4.1b}$$

$$\frac{\partial \mathbf{v}}{\partial t} + \bar{\mathbf{v}} \cdot \nabla \mathbf{v} - \nabla \theta + y\mathbf{v}^\perp = -\mathbf{v} \cdot \nabla \bar{\mathbf{v}}, \tag{4.1c}$$

$$\frac{\partial \theta}{\partial t} + \bar{\mathbf{v}} \cdot \nabla \theta - \nabla \cdot \mathbf{v} = \delta^2 (\bar{H}a - S^\theta), \tag{4.1d}$$

$$\frac{\partial q}{\partial t} + \bar{\mathbf{v}} \cdot \nabla q + \tilde{Q} \nabla \cdot \mathbf{v} = -\delta^2 (\bar{H}a - S^q), \tag{4.1e}$$

$$\frac{\partial a}{\partial t} = \Gamma q a. \tag{4.1f}$$

These equations combine the MJO skeleton model (Majda and Stechmann 2009b) and nonlinear interactions between the baroclinic and barotropic modes (Majda

Table 4.1 Parameters of the MJO skeleton model, and relation to small parameter δ

Par.	Non-dim. val.	Dim. val.	Description
Γ	1	$\sim$0.5/day/(g/kg)	Convective growth/decay rate
$\bar{H}$	0.23	$\sim$10 K/day	Parameter to rescale a
$\delta^2\bar{a}$	δ^2		Convective activity envelope at RCE state
$\tilde{Q}$	0.9		Non-dim. background vertical moisture gradient
$\delta^2 S^\theta$	$\delta^2\bar{H}$		Radiative cooling rate
$\delta^2 S^q$	$\delta^2\bar{H}$		Moisture source

and Biello 2003). The equations have been nondimensionalized using standard equatorial reference scales (Chen et al. 2015). Here $\bar{\mathbf{v}} = (\bar{u}, \bar{v})$ and $\bar{p}$ are barotropic velocity and pressure; $\mathbf{v} = (u, v)$ and θ are baroclinic velocity and potential temperature; and q is water vapor (sometimes referred to as "moisture"). The tropical convective activity envelope is denoted by $\delta^2 a$, where δ is a small parameter that monitors the scales of tropical convection envelope. Likewise, δ^2 is also applied to quantities S^θ and S^q, radiative cooling, and the moisture source. In this paper, $\delta^2 S^\theta$ and $\delta^2 S^q$ are set to be constants for energy conservation, although they usually have both spatial and temporal variance in reality. Together with Γ, $\bar{H}$, and $\tilde{Q}$, the coefficients are described in Table 4.1.

In Eqs. (4.1), $\mathbf{v}^\perp = (-v, u)$ is the β-plane approximation of tropical Coriolis force, and (x, y) denotes the zonal and meridional directions. Without the moisture q and convection envelope a, system (4.1) is the two-vertical-mode Galerkin truncation for the Boussinesq equations with rigid-lid boundary on the vertical direction (see e.g., Neelin and Zeng (2000); Majda and Shefter (2001); Majda and Biello (2003); Khouider and Majda (2005)). Without the barotropic wind $\bar{\mathbf{v}}$, the system is the MJO skeleton model on the first baroclinic mode (see e.g., Majda and Stechmann (2009b, 2011)).

Note that for the primitive equations (4.1), a total energy is conserved, and it is composed of four parts, dry barotropic energy $\mathscr{E}_T$, dry baroclinic energy $\mathscr{E}_C$, moisture energy $\mathscr{E}_M$, and convective energy $\mathscr{E}_A$:

$$\mathscr{E}_T(t) = \frac{1}{2} \int_{-Y}^{Y} \int_0^X |\bar{\mathbf{v}}|^2 \mathrm{d}x\mathrm{d}y \tag{4.2a}$$

$$\mathscr{E}_C(t) = \frac{1}{4} \int_{-Y}^{Y} \int_0^X |\mathbf{v}|^2 + \theta^2 \mathrm{d}x\mathrm{d}y \tag{4.2b}$$

$$\mathscr{E}_M(t) = \frac{1}{4} \int_{-Y}^{Y} \int_0^X \frac{1}{\tilde{Q}(1-\tilde{Q})}(q + \tilde{Q}\theta)^2 \mathrm{d}x\mathrm{d}y \tag{4.2c}$$

$$\mathscr{E}_A(t) = \frac{\delta^2}{2} \int_{-Y}^{Y} \int_0^X \frac{1}{\tilde{Q}\Gamma}\left[\bar{H}a - S^\theta \log(a)\right] \mathrm{d}x\mathrm{d}y \tag{4.2d}$$

The individual components of this energy had previously been identified separately in Majda and Stechmann (2009b) and Majda and Biello (2003), and in the coupled system the components are combined together. Note that in (4.1f), the conserved quantity $\mathscr{E}$ still holds if a is also advected by the barotropic wind, i.e., with an additional term $\bar{u} \cdot \nabla a$ in (4.1f). However, we do not include this term here because later, when meridional truncation is applied to the system, the energy conservation will not hold with this additional term.

Using the streamfunction ψ for barotropic mode, which satisfies $(\bar{u}, \bar{v}) = (-\psi_y, \psi_x)$, the barotropic equation can also be written as

$$\frac{D}{Dt}\Delta\psi + \psi_x + \frac{1}{2}\nabla \cdot \left[-(vu)_y + (vv)_x\right] = 0, \tag{4.3}$$

where

$$\frac{D}{Dt} = \frac{\partial}{\partial t} + \bar{u}\frac{\partial}{\partial x} + \bar{v}\frac{\partial}{\partial y}$$

represents advection by the barotropic wind.

4.2.2   The Zonal-Long-Wave-Scaled Model

To consider the planetary scale of the coupled equations in (4.3) and (4.1c)–(4.1f), zonal variations are assumed to depend on a longer scale, as are temporal variations. The long zonal and long temporal coordinates are introduced as

$$x' = \delta x, \qquad t' = \delta t. \tag{4.4}$$

Correspondingly, the meridional velocity is also scaled so that

$$v = \delta v'. \tag{4.5}$$

The equations in (4.3) and (4.1c)–(4.1f) become the long-wave-scaled system:

$$\frac{D}{Dt'}\psi_{yy} + \psi_{x'} - \frac{1}{2}\nabla' \cdot \left[(v'u)_y\right] + \delta^2\left\{\frac{D}{Dt'}\psi_{x'x'} + \frac{1}{2}\nabla' \cdot \left[(v'v')_{x'}\right]\right\} = 0, \tag{4.6a}$$

$$\frac{D}{Dt'}u - \theta_x - yu - \mathbf{v}' \cdot \nabla'\psi_y = 0, \tag{4.6b}$$

$$-\theta_y + yu + \delta^2\left(\frac{D}{Dt'}v' + v' \cdot \nabla'\psi_{x'}\right) = 0 \tag{4.6c}$$

$$\frac{D}{Dt'}\theta - \nabla' \cdot \mathbf{v}' - \delta(\bar{H}a - S^\theta) = 0, \tag{4.6d}$$

$$\frac{D}{Dt'}q + \tilde{Q}\nabla' \cdot \mathbf{v}' + \delta(\bar{H}a - S^q) = 0, \tag{4.6e}$$

$$\frac{\partial}{\partial t'}a - \Gamma qa = 0. \tag{4.6f}$$

Here the primes represent the long-wave scaled coordinates and variables:

$$\frac{D}{Dt'} = \frac{\partial}{\partial t'} + u\frac{\partial}{\partial x'} + v'\frac{\partial}{\partial y'}, \quad \nabla' = (\frac{\partial}{\partial x'}, \frac{\partial}{\partial y}), \quad \text{and} \quad \bar{\mathbf{v}}' = (u, v').$$

Note that, if the moisture and convection are neglected, then system (4.6) would be in the same form as in Majda and Biello (2003).

Furthermore, the convective activity a will be written as an anomaly from the state of radiative-convective equilibrium (RCE):

$$S^\theta = S^q = \bar{H}\bar{a},$$

where these constants take the values given in Table 4.1. When $a = \bar{a}$, the external radiative cooling/moisture source is in balance with the source/sink from the convection, and the system achieves RCE. By writing a as an anomaly with respect to the RCE value $\bar{a}$, the forcing terms in (4.6d) and (4.6e) can be written as

$$\bar{H}a - S^\theta = \bar{H}a - S^q = \bar{H}(a - \bar{a}) = \bar{H}a', \tag{4.7}$$

and the equation for the convection envelope (4.6f) can be written as

$$\frac{\partial}{\partial t'}a' - \Gamma\bar{a}q = \Gamma a'q. \tag{4.8}$$

4.2.3   Long Time Scales for Tropical-Extratropical Interactions

The small parameter δ is also used for introducing two longer time scales:

$$T_1 = \delta t', \quad T_2 = \delta^2 t',$$

which match the units of intraseasonal timescales as appeared in Majda and Biello (2003) and Majda and Stechmann (2009a).

4.2.4 Asymptotic Expansions

The solutions are assumed to have an asymptotic structure with **ansatz**:

$$(\psi, u, v', \theta, q) = \delta^2(\psi_1, u_1, v_1, \theta_1, q_1) + \delta^3(\psi_2, u_2, v_2, \theta_2, q_2)$$

$$+ \delta^4(\psi_3, u_3, v_3, \theta_3, q_3) + o(\delta^4) \tag{4.9a}$$

and $$a' = \delta a_1 + \delta^2 a_2 + \delta^3 a_3 + O(\delta^4). \tag{4.9b}$$

This small amplitude assumption is consistent with a small Froude number assumption as in Majda and Biello (2003). All variables are assumed to depend on three time scales: t' and $T_1 = \delta t'$ and $T_2 = \delta^2 t'$.

4.2.5 Eigenmodes of the Linear System

In this section, the eigenmodes are described for the baroclinic system and barotropic system. (In subsequent sections, these linear eigenmodes will be used in the identification of three-wave resonances.)

First, for the baroclinic part $(\mathbf{v}, \theta, q, a)$ of the system (4.6), its basic four eigenmodes have been described previously (Majda and Stechmann 2009b): dry Kelvin, MJO, moist Rossby, and dry Rossby, as shown in Fig. 4.1. In brief, the dry Kelvin wave is a fast eastward-propagating wave; the MJO is a slow east-propagating wave; and the moist and dry baroclinic Rossby waves are slow and fast westward-propagating waves, respectively. Notice that the baroclinic Rossby and Kelvin waves in this model are dry equatorial waves, not convectively coupled equatorial waves (CCEWs) (Kiladis et al. 2009). The model here is a planetary-scale model that does not explicitly resolve CCEWs, which occur mainly on smaller,

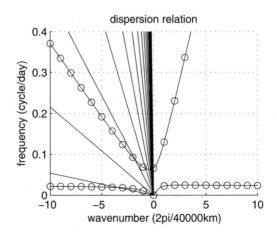

Fig. 4.1 Dispersion relation of the KRQA system (with circles) and barotropic wind with different meridional wavelengths (without circles)

synoptic scales. Nevertheless, one would expect similar wave interactions to hold for CCEWs, due to the similar structures of dry and convectively coupled waves, if a model were used that resolved such synoptic-scale convectively coupled waves.

Second, for the barotropic part (ψ) of the system (4.6), the dispersion relation is written explicitly as

$$\omega_T(k) = -\frac{k}{L^2}. \tag{4.10}$$

The above formula implies that barotropic Rossby waves travel faster for smaller meridional wavenumber L or, equivalently, longer meridional wavelength Y. In Fig. 4.1, barotropic dispersion relation are shown with different meridional wavenumbers.

4.2.6 The Reduced Asymptotic Models

In this section, the reduced asymptotic model is described and it demonstrates the resonant interactions among MJO, Kelvin and extratropical waves. The reduced asymptotic model is an ODE system and it has energy conservation which is analogous to the energy conservation of the whole PDE system.

4.2.6.1 Derivation

The derivation of (4.15) starts by assuming a leading order barotropic component of the form

$$\psi_1 \Leftrightarrow \beta(T_1, T_2)e^{i\theta_T} + \text{C.C.}, \tag{4.11}$$

where $\theta_T = k_T x' + \omega_T t'$, with k_T and ω_T the wavenumber and frequency, respectively, of the barotropic Rossby wave, C.C. stands for complex conjugate, and leading order baroclinic variables of the form

$$(u_1, v_1, \theta_1, q_1, a_1) \Leftrightarrow \alpha_{j_1} e^{i\theta_{j_1}} \mathbf{r}_{j_1} + \alpha_{j_2} e^{i\theta_{j_2}} \mathbf{r}_{j_2} + \text{C.C.}, \tag{4.12}$$

where $\theta_j = k_j x' + \omega_j t'$, and $\mathbf{r}$ represents a right eigenvector of the linear system, and it is normalized with respect to the energy as

$$E_j = \mathbf{r}_j^{\dagger} \mathcal{H} \mathbf{r}_j = 1, \quad j = j_1, j_2, \tag{4.13}$$

where $\mathcal{H}$ is the Hessian matrix of the conserved energy for the linear baroclinic system.

Next is the key step of multi-scale analysis. With plugging (4.11) and (4.12) into the expanded systems and applying multiscale asymptotic analysis, three-wave triads are identified when they satisfy the following resonance conditions:

$$k_{j_1} + k_{j_2} + k_T = 0, \tag{4.14a}$$

$$\omega_{j_1} + \omega_{j_2} + \omega_T = 0. \tag{4.14b}$$

Further details of the derivation can be found in Chen et al. (2015). The result is the following reduced asymptotic model:

$$-\partial_{T_2}\beta \qquad\qquad + id_2\beta + id_3\alpha_{j_1}{}^*\alpha_{j_2}{}^* = 0, \tag{4.15a}$$

$$-\partial_{T_2}\alpha_{j_1} + id_4\alpha_{j_1}{}^2\alpha_{j_1}{}^* + id_5\alpha_{j_1} + id_6\beta^*\alpha_{j_2}{}^* = 0, \tag{4.15b}$$

$$-\partial_{T_2}\alpha_{j_2} + id_7\alpha_{j_2}{}^2\alpha_{j_2}{}^* + id_8\alpha_{j_2} + id_9\beta^*\alpha_{j_1}{}^* = 0, \tag{4.15c}$$

where the coefficients d_j are all real numbers. This is a system of ODEs for three-wave resonance, where β is the complex amplitude of the barotropic Rossby waves, and α_{j1} and α_{j2} are the complex amplitudes of two baroclinic waves. For the applications of importance here, one of the αs will correspond to the MJO, and this ODE system describes its interaction with the other waves, with an aim toward MJO initiation and termination.

In brief, the terms in (4.15) fall into three groups that are linked to different features in the full 2D model (4.1): the cubic terms are associated with the nonlinear q-a interactions; the linear terms come from dispersive terms; the quadratic terms arise from the nonlinear baroclinic-barotropic interactions. Here, in particular, the cubic self-interaction terms are a novel feature that are not typically found in three-wave resonance ODEs, and they arise here from the effects of water vapor q and convective activity a.

4.2.6.2 Energy Conservation

The ODE system (4.15) satisfies the following energy conservation principle:

$$\frac{\partial}{\partial T_2}(\beta\beta^* + \alpha_{j_1}\alpha_{j_1}{}^* + \alpha_{j_2}\alpha_{j_2}{}^*) = 0. \tag{4.16}$$

To see that this quantity is conserved, one must use the fact that all coefficients of the ODE system are purely imaginary and satisfy the condition

$$d_3 + d_6 + d_9 = 0, \tag{4.17}$$

which is a key component for energy conservation in three-wave interaction equations (Majda 2003).

4.2.7   Tropical–Extratropical Interactions via Three-Wave Resonant Interactions

With the ODE system (4.15) in hand, it can now be used to investigate the nature of tropical–extratropical interactions via three-wave resonant interactions.

To start, one must consider which waves satisfy the resonance conditions. Figure 4.2 shows the resonance condition for three-wave interactions between MJO, dry baroclinic Kelvin, and barotropic Rossby waves. This figure identifies a particular resonant interaction where α_{j_1} and α_{j_2} will be the amplitudes for the MJO and dry baroclinic Kelvin waves, respectively.

To consider a case of MJO initiation, the initial conditions are $\alpha_{j_1} = 0$ and $|\alpha_{j_2 (t=0)}| = |\beta_{(t=0)}| = 1$, and the evolution of the ODEs is shown in Fig. 4.3. The waves exchange energy periodically, with period of about 120 days, roughly the time scale for MJO initiation and termination in nature. At early times, the MJO amplitude grows (i.e., the MJO initiates) by extracting energy from the other two waves, mainly from the Kelvin wave. Termination of the MJO then follows from a transfer of energy back to the Kelvin wave. It is interesting to note that the MJO extracts more energy from the (tropical) Kelvin wave than it does from the (extratropical) barotropic Rossby wave. Hence, this case does not clearly represent an extratropical route to MJO initiation and termination. Nevertheless, the extratropical Rossby wave plays a key role, since its presence is needed to facilitate the three-wave interactions.

Fig. 4.2 Resonance condition for three-wave interactions, where circles indicate the particular wavenumbers and frequencies that lead to resonances for interaction of the MJO, Kelvin, and barotropic Rossby wave

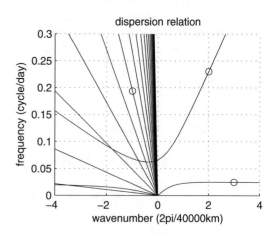

Fig. 4.3
MJO–Kelvin–barotropic
Rossby wave interactions.
Initial condition: $\alpha_{j_1} = 0$,
$|\alpha_{j_2(t=0)}| = |\beta_{(t=0)}| = 1$

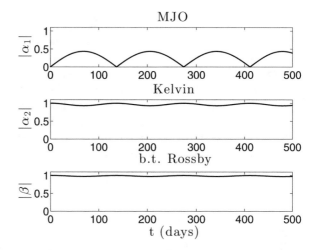

4.3 Direct Two-Wave Interactions with Climatological Mean Flow

Earlier, in Sect. 4.2, three-wave interactions were investigated, and it was seen that the MJO extracted only a small amount of energy from the (extratropical) barotropic Rossby waves. If not via three-wave interactions, then in what way can the MJO exchange significant energy with extratropical waves? To explore an alternative scenario, as motivated by earlier studies described in the introduction section, we now consider the impact of the Walker circulation, which is a climatological mean circulation in the tropics (Gill 1980; Stechmann and Ogrosky 2014). The Walker circulation introduces zonal (x) variations into the system, which were not present in the zonally homogeneous setup of Sect. 4.2. It occurs in the skeleton model (4.1) by including unbalanced moisture source and radiative cooling, i.e., $S^q \neq S^\theta$. In what follows, in the presence of the Walker circulation, we will see that the MJO and barotropic Rossby waves interact directly with significant energy exchanges.

For the moisture source and radiative cooling, it is assumed that

$$S^q = \overline{S^q} + \delta S^q{}_1, \qquad S^\theta = \overline{S^\theta} + \delta S^\theta{}_1, \qquad (4.18)$$

where $\overline{\cdot} = \int \cdot \, dx dy$ is the mean value over the horizontal domain. In the presence of the Walker circulation, the energy conservation no longer holds. The Walker circulation here behaves as an energy source/sink for the MJO mode and the barotropic Rossby wave.

We take an MJO linear mode as the leading order in the ansatz for the baroclinic system, and we choose the barotropic Rossby mode such that their wavenumbers and frequencies satisfy the resonance conditions in the presence of the Walker circulation:

$$k_{\text{MJO}} + k_{\text{W}} + k_{\text{T}} = 0, \tag{4.19a}$$

$$\omega_{\text{MJO}} + \omega_{\text{T}} = 0, \tag{4.19b}$$

where k_{MJO}, k_{W}, and k_{T} are the wavenumbers for the MJO, the Walker circulation, and the barotropic Rossby wave, and ω_{MJO} and ω_{T} are the wave frequencies for the MJO and the barotropic Rossby wave. The frequency for the Walker circulation, ω_{W}, is zero. This type of resonant condition is analogous to topographic resonance (Majda et al. 1999). Because MJO and barotropic Rossby waves travel in opposite directions, the wavenumber of the Walker circulation has to satisfy the condition

$$|k_{\text{W}}| \geq 2$$

to ensure that both the MJO and barotropic Rossby waves have at least 1 complete cycle along the equator.

To carry out the multiscale analysis, we write the leading order baroclinic solution as

$$(u_1, v_1, \theta_1, q_1, a_1) \Leftrightarrow \alpha(T_1, T_2)e^{i(k_{\text{MJO}}x - \omega_{\text{MJO}}t)}\mathbf{r}_{\text{MJO}} + \mathbf{r}_{\text{W}} + \text{C.C.}. \tag{4.20}$$

It is the sum of the MJO mode and the Walker circulation. The leading order barotropic ansatz is

$$\psi_1 \Leftrightarrow \frac{1}{\sqrt{2\pi L}}\beta(T_1, T_2)e^{i(k_{\text{T}}x - \omega_{\text{T}}t)} + \text{C.C.}, \tag{4.21}$$

where C.C. stands for the complex conjugates, and $\mathbf{r}_{\text{MJO}}$ is the right eigenvector for the MJO mode. The eigenvector for the MJO mode is normalized by the baroclinic energy.

Following similar systematic asymptotic procedures as in Chen et al. (2015) leads to a reduced ODE model for the amplitudes for the linear modes:

$$\partial_{T_2}\beta \qquad\qquad + id_2\beta + h_3\alpha^* = 0, \tag{4.22a}$$

$$\partial_{T_2}\alpha + id_4\alpha^2\alpha^* + id_5\alpha + h_6\beta^* = 0. \tag{4.22b}$$

The energy exchanges in this ODE model are illustrated in Fig. 4.4. The numerical simulation shows that the MJO initially gains energy and the barotropic Rossby wave loses energy, and the total energy is increasing until it peaks at around 70 days. After this time, the MJO mode decays in amplitude as the barotropic Rossby wave gains energy and returns to the original state. This patterns repeats itself as a nonlinear cycle with time period of roughly 140 days.

To illustrate the spatial variations, Fig. 4.5 shows the Hovmoller diagram for $\bar{H}a_{1a}$, the leading order anomaly of the convective activity. In this figure, the MJO is traveling eastward at a speed of 5 m/s, and the wave amplitude is zero at 0 day, peaks at around 70 days, and returns to zero-amplitude at 140 days. This corresponds to

Fig. 4.4 MJO initiation in
the presence of a Walker
circulation. $k_{\mathrm{MJO}} = 1$,
$k_{\mathrm{W}} = -2$, and $k_{\mathrm{T}} = 1$

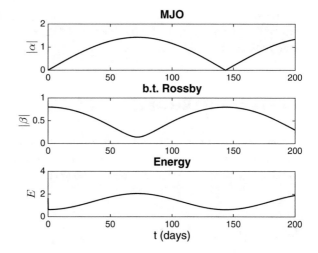

Fig. 4.5 Hovmoller diagram
of $\bar{H} a_1$ for MJO initiation.
Solid contours denote positive
anomaly; dashed contours
denote negative anomaly. The
contour interval is 0.0016 K/d

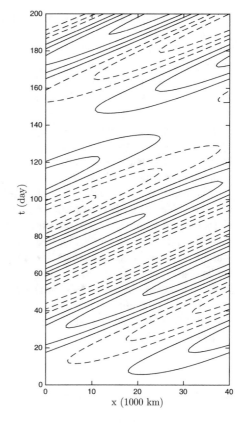

a wave train of roughly one or two MJO events, depending on the spatial location, similar to the organization of sequences of MJO events in nature (Thual et al. 2014; Yoneyama et al. 2013).

4.4   Additional Realism

In this section, two additional features are explored to add more realism to the interactions between the MJO and extratropics: (1) a more general Walker circulation with multiple wavenumbers, and (2) the effects of wind shear.

4.4.1   More general Walker circulation

In the previous section, the case for the sinusoidal Walker circulation with wavenumber $k_W = 2$ is discussed. The realistic Walker circulation, on the other hand, is composed of a variety of wavenumbers. For example, Ogrosky and Stechmann (2015) described simplified versions of the Walker circulation using 1 or 3 Fourier modes in their study. In this section, another mode for the Walker cell, $k_W = 3$ is also included in addition to $k_W = 2$. In this case, Fig. 4.6 shows a Hovmoller diagram for the convective envelope of the leading order MJO waves with wavenumbers 1 and 2. A wave packet is presented in the diagram with a life cycle around 150 days. These cases illustrate additional realism, such as MJOs with more realistic zonal variations, through the interaction with the Walker circulation with more realistic zonal variations.

It is interesting to note that the MJO skeleton has acquired, in Figs. 4.5 and 4.6, a group velocity that is westward. The westward group velocity is a consequence of interaction with the Walker circulation, as also seen earlier in Chap. 2 and seen by Majda and Stechmann (2011); in contrast, for a homogeneous background state without a Walker circulation, the MJO skeleton eigenmode has a group velocity that is eastward (see Chap. 2 and Majda and Stechmann (2009b, 2011)). A westward group velocity of observed MJOs has been suggested (see Adames and Kim (2016) and references therein), and the present results suggest the possibility that a zonally varying climatological mean state may play a role in determining the group velocity of the observed MJO.

4.5   Effects of Wind Shear

This section includes the effect of the horizontal and vertical wind shear in the model. In the past, dry models (Hoskins and Jin 1991; Majda and Biello 2003; Webster 1972, 1981, 1982) suggest that wind shear can significantly affect energy

Fig. 4.6 Hovmoller diagram of $\bar{H}a_1$ for MJO initiation with two MJO modes: $k_{\mathrm{MJO1}} = 1$ and $k_{\mathrm{MJO2}} = 2$. Solid contours denote positive anomaly; dashed contours denote negative anomaly. The contour interval is 0.0053 K/d

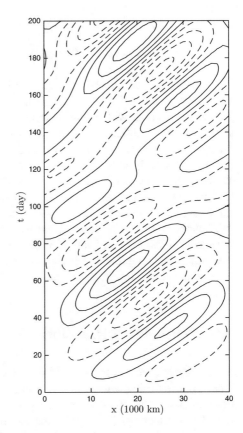

transfer between the barotropic waves and tropical waves. Both zonally uniform and zonally varying wind shear have been considered in these past studies. However, the MJO, water vapor, and convective activity were not explicitly included in these models. Motivated by previous studies, here we consider both barotropic and baroclinic wind shear that is $O(\delta^2)$:

$$(\tilde{u}(x, y, z), \tilde{v}(x, y, z), \tilde{w}(x, y, z)) = \left(\tilde{U}(y, z), 0, 0\right), \tag{4.23}$$

where

$$\tilde{U}(y, z) = \delta^2 \left[U_0 + L\sin(Ly)B_0 + \cos(\pi z)\left(u^{(0)}{}_0\Phi_0 + u^{(2)}{}_0\Phi_2\right)\right]. \tag{4.24}$$

Here U_0 is the constant global mean flow, B_0 is the strength of the barotropic wind shear on the meridional direction, and $u^{(0)}{}_0$, $u^{(2)}{}_0$ are the strengths of the baroclinic wind shear on both vertical and meridional directions. A similar multi-scale analysis is carried out, and the resonance condition is not affected by the wind shear (Chen et al. 2016). The reduced ODE model for the MJO–barotropic Rossby wave interaction is:

$$\partial_{T_2}\beta \qquad\qquad\qquad + i(d_2 + f_1)\beta + h_3\alpha^* = 0, \qquad\qquad (4.25a)$$

$$\partial_{T_2}\alpha + i d_4\alpha^2\alpha^* + i(d_5 + f_2)\alpha + h_6\beta^* = 0. \qquad\qquad (4.25b)$$

The wind shear introduces two additional linear terms with coefficients f_1 and f_2, both of which are real values. (The resonance condition is the same as it was earlier in the case without wind shear.) Numerical simulations were performed for MJO initiation with the effects of barotropic shear. Four different barotropic shear profiles were considered. (1) $U_0 = 0$, $B_0 = 1$, (2) $U_0 = -1$, $B_0 = 1$, (3) $U_0 = 1$, $B_0 = 0$, and (4) $U_0 = 1$, $B_0 = 1$. The results for the four cases (not shown here) suggest that the barotropic shear has little effect on the maximum amplitude attained by the MJO and the time period of the solution. The small effect of wind shear here differs from the important effects of wind shear seen by Majda and Biello (2003), and several differences in the models could contribute to the different results. In particular, in the model of Majda and Biello (2003), the waves are all nondispersive; resonant interactions are included for all wavenumbers rather than a small discrete subset of wavenumbers; and moisture and convection are not explicitly included.

4.6 Concluding Discussions

In this chapter, models for tropical–extratropical interactions of the MJO skeleton were explored using multiscale asymptotic analysis. First, with a zonally homogeneous base state, three-wave interactions were investigated. In this case, with the MJO interacting with Kelvin and barotropic Rossby waves, it was seen that the main interactions appear between the MJO and Kelvin wave. The barotropic Rossby wave plays an important role as its presence is required for the MJO and Kelvin wave to exchange energy, but the barotropic Rossby wave does not itself exchange significant energy with the MJO directly. Secondly, a case was explored with a Walker circulation, which provides a zonally varying climatological mean state. In this case, the MJO and the extratropical Rossby waves interact directly and exchange significant energy.

Finally, in the presence of the Walker circulation, two additional features were explored to include additional realism: the effect of a more general Walker circulation, which creates a more realistic MJO wave packet, and the effect of wind shear, which does not appear to affect the MJO–extratropical wave interactions.

In the future, it would be interesting to extend these results to include seasonal variations with an aim toward the boreal summer intraseasonal oscillation and monsoon intraseasonal oscillations (Lawrence and Webster 2002; Wheeler and Hendon 2004; Lau and Waliser 2012).

References

Adames ÁF, Kim D (2016) The MJO as a dispersive, convectively coupled moisture wave: theory and observations. J Atmos Sci 73(3):913–941

Chen S, Majda AJ, Stechmann SN (2015) Multiscale asymptotics for the skeleton of the Madden–Julian oscillation and tropical–extratropical interactions. Math Clim Weather Forecast 1:43–69. https://doi.org/10.1515/mcwf-2015-0003

Chen S, Majda AJ, Stechmann SN (2016) Tropical–extratropical interactions with the MJO skeleton and climatological mean flow. J Atmos Sci 73(10):4101–4116. https://doi.org/10.1175/JAS-D-16-0041.1

Ferranti L, Palmer TN, Molteni F, Klinker E (1990) Tropical-extratropical interaction associated with the 30–60 day oscillation and its impact on medium and extended range prediction. J Atmos Sci 47(18):2177–2199

Frederiksen JS, Frederiksen CS (1993) Monsoon disturbances, intraseasonal oscillations, teleconnection patterns, blocking, and storm tracks of the global atmosphere during January 1979: linear theory. J Atmos Sci 50(10):1349–1372

Gill AE (1980) Some simple solutions for heat-induced tropical circulation. Q J Roy Meteorol Soc 106(449):447–462

Gloeckler LC, Roundy PE (2013) Modulation of the extratropical circulation by combined activity of the Madden–Julian Oscillation and equatorial Rossby waves during boreal winter. Mon Weather Rev 141(4):1347–1357

Hoskins BJ, Ambrizzi T (1993) Rossby wave propagation on a realistic longitudinally varying flow. J Atmos Sci 50(12):1661–1671

Hoskins BJ, Jin FF (1991) The initial value problem for tropical perturbations to a baroclinic atmosphere. Q J Roy Meteorol Soc 117(498):299–317

Jin F, Hoskins BJ (1995) The direct response to tropical heating in a baroclinic atmosphere. J Atmos Sci 52(3):307–319

Khouider B, Majda AJ (2005) A non-oscillatory balanced scheme for an idealized tropical climate model: part I: algorithm and validation. Theor Comput Fluid Dyn 19(5):331–354

Kiladis GN, Wheeler MC, Haertel PT, Straub KH, Roundy PE (2009) Convectively coupled equatorial waves. Rev Geophys 47:RG2003. https://doi.org/10.1029/2008RG000266

Lau WKM, Waliser DE (eds) (2012) Intraseasonal variability in the atmosphere–ocean climate system, 2nd edn. Springer, Berlin

Lawrence DM, Webster PJ (2002) The boreal summer intraseasonal oscillation: relationship between northward and eastward movement of convection. J Atmos Sci 59(9):1593–1606

Liebmann B, Hartmann DL (1984) An observational study of tropical–midlatitude interaction on intraseasonal time scales during winter. J Atmos Sci 41(23):3333–3350

Lin H, Brunet G, Derome J (2009) An observed connection between the North Atlantic Oscillation and the Madden–Julian oscillation. J Clim 22(2):364–380

Majda AJ (2003) Introduction to PDEs and waves for the atmosphere and ocean. Courant lecture notes in mathematics, vol 9. American Mathematical Society, Providence

Majda AJ, Biello JA (2003) The nonlinear interaction of barotropic and equatorial baroclinic Rossby waves. J Atmos Sci 60:1809–1821

Majda AJ, Shefter MG (2001) Models of stratiform instability and convectively coupled waves. J Atmos Sci 58:1567–1584

Majda AJ, Stechmann SN (2009a) A simple dynamical model with features of convective momentum transport. J Atmos Sci 66:373–392

Majda AJ, Stechmann SN (2009b) The skeleton of tropical intraseasonal oscillations. Proc Natl Acad Sci USA 106(21):8417–8422

Majda AJ, Stechmann SN (2011) Nonlinear dynamics and regional variations in the MJO skeleton. J Atmos Sci 68:3053–3071

Majda AJ, Rosales RR, Tabak EG, Turner CV (1999) Interaction of large-scale equatorial waves and dispersion of Kelvin waves through topographic resonances. J Atmos Sci 56(24):4118–4133

Matthews AJ, Kiladis GN (1999) The tropical–extratropical interaction between high-frequency transients and the Madden–Julian oscillation. Mon Weather Rev 127(5):661–677

Matthews AJ, Hoskins BJ, Masutani M (2004) The global response to tropical heating in the Madden–Julian oscillation during the northern winter. Q J Roy Meterol Soc 130(601):1991–2011

Neelin JD, Zeng N (2000) A quasi-equilibrium tropical circulation model—formulation. J Atmos Sci 57:1741–1766

Ogrosky HR, Stechmann SN (2015) The MJO skeleton model with observation-based background state and forcing. Q J Roy Meteorol Soc 141(692):2654–2669. https://doi.org/10.1002/qj.2552

Ray P, Zhang C (2010) A case study of the mechanics of extratropical influence on the initiation of the Madden–Julian oscillation. J Atmos Sci 67(2):515–528

Roundy PE (2011) Tropical–extratropical interactions. In: Lau WKM, Waliser DE (eds) Intraseasonal variability in the atmosphere–ocean climate system. Springer, Berlin

Stechmann SN, Ogrosky HR (2014) The Walker circulation, diabatic heating, and outgoing longwave radiation. Geophys Res Lett 41:9097–9105. https://doi.org/10.1002/2014GL062257

Thual S, Majda AJ, Stechmann SN (2014) A stochastic skeleton model for the MJO. J Atmos Sci 71:697–715

Webster PJ (1972) Response of the tropical atmosphere to local, steady forcing. Mon Weather Rev 100(7):518–541

Webster PJ (1981) Mechanisms determining the atmospheric response to sea surface temperature anomalies. J Atmos Sci 38(3):554–571

Webster PJ (1982) Seasonality in the local and remote atmospheric response to sea surface temperature anomalies. J Atmos Sci 39(1):41–52

Weickmann KM (1983) Intraseasonal circulation and outgoing longwave radiation modes during Northern Hemisphere winter. Mon Weather Rev 111(9):1838–1858

Weickmann K, Berry E (2009) The tropical Madden–Julian oscillation and the global wind oscillation. Mon Weather Rev 137(5):1601–1614

Weickmann KM, Lussky GR, Kutzbach JE (1985) Intraseasonal (30–60 day) fluctuations of outgoing longwave radiation and 250 mb streamfunction during northern winter. Mon Weather Rev 113(6):941–961

Wheeler MC, Hendon HH (2004) An all-season real-time multivariate MJO index: development of an index for monitoring and prediction. Mon Weather Rev 132(8):1917–1932

Yoneyama K, Zhang C, Long CN (2013) Tracking pulses of the Madden-Julian oscillation. Bull Am Meteorol Soc 94(12):1871–1891

Chapter 5
New Indices for Observations of Tropical Variability Based on the Skeleton Model and a Model for the Walker Circulation

How can the Madden–Julian oscillation (MJO) be identified in observational data? Some methods identify the MJO using wind data only, other methods use cloudiness or precipitation data only, and others use a combination of winds and cloudiness. In a sense, each MJO index offers a different definition of what the MJO is. While MJO indices have traditionally been *empirical*, the MJO skeleton model offers the possibility of a *theoretical* definition of the MJO. This chapter describes MJO skeleton indices based on the skeleton model's theoretical prediction of the MJO's structure (as introduced earlier in Chap. 2). In addition to their practical use in monitoring MJO activity, the indices allow further tests and comparisons of theory and observational data: How good is the statistical agreement between the observed MJO and stochastic skeleton model simulations (introduced earlier in Chap. 3)? Furthermore, similar methodologies can be used to model other aspects of tropical weather and climate, such as the Walker circulation and its variability associated with the El Niño–Southern oscillation (ENSO).

5.1 Introduction

In earlier chapters, the skeleton model was introduced as a model for the dominant component of tropical intraseasonal variability, namely the MJO. The model consists of a system of nonlinear PDEs that couple the dry dynamics of the tropical troposphere to lower-tropospheric moisture and deep convective heating. Solutions to the skeleton model were shown to exhibit several features of observed variability, including a coupled convection-circulation structure that appears in composites of observed MJOs (Hendon and Salby 1994).

The primary goal of the current chapter is to extend this comparison with observations by assessing just how prominently the model's features appear in observational data. While several data analysis methods could be used for this task,

© The Author(s), under exclusive licence to Springer Nature Switzerland AG 2019   67
A. J. Majda et al., *Tropical Intraseasonal Variability and the Stochastic Skeleton Method*, SpringerBriefs in Mathematics of Planet Earth, https://doi.org/10.1007/978-3-030-22247-5_5

there are two possible approaches that merit immediate comment. First, one might look for signals in the data which propagate with the theoretical group and phase velocities; this method utilizes the *eigenvalues* of the model's solutions. Methods based on this approach have been used to identify the MJO, convectively coupled equatorial waves, and other phenomena in the tropical atmosphere (e.g., Wheeler and Kiladis 1999; Wheeler et al. 2000; Wheeler and Weickmann 2001).

A second possibility is to look for signals in the data which exhibit the expected structure; such a method utilizes the *eigenvectors* of the theory. This is the approach used here. There are several reasons to pursue such a method. For one, the method provides a clear way to look for signals in multivariate data. For another, by avoiding the use of temporal filtering, the method may be used to identify signals in real time.

The data analysis technique developed here will result in an index that describes how clearly the skeleton model's structure appears in observational data on any given day. A brief comparison with other methods based on space–time filtering and empirical orthogonal functions (EOFs) will be made. In addition, it will be shown that this technique has other applications, including identifying the strength of the Walker circulation, a primary component of the tropical atmosphere's background state. Since the Walker circulation may be modeled by a subset of the skeleton model equations, it provides an ideal setting for introducing the data analysis methods used here. Other applications will be discussed as well, including an assessment of the length and time scales on which the equatorial long-wave approximation is observed to be valid.

The rest of this chapter is thus organized as follows: we will first discuss a simple model of the Walker circulation and introduce the data analysis techniques that will be used throughout the chapter. These methods show that the Walker circulation may surprisingly be modeled without reliance on damping terms commonly used. We then study linear solutions to the skeleton model, first with a zonally uniform background state, and then with a zonally varying background state. The signals of these solutions are found by projecting observational data onto the model's eigenvectors. We will conclude by using these methods to assess the degree to which the tropical atmosphere exhibits the equatorial long-wave dynamics that is a component of the models discussed here.

5.2   Data Analysis and Modeling of the Walker Circulation

The Walker circulation is an east–west overturning of air in the tropical atmosphere. This overturning is coupled with cloudiness and precipitation; ascending air occurs in regions of deep convection and high precipitation rates and descending air is correlated with desert regions (see Fig. 5.1a for a schematic). As a strong component of the climate system, the Walker circulation has global significance and is closely

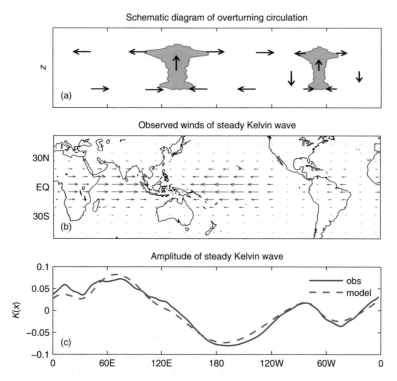

Fig. 5.1 (**a**) Schematic diagram of east–west overturning circulations near the equator. (**b**) The zonal winds of the Kelvin wave as observed in NCEP/NCAR reanalysis data, averaged from 1980 to 2009. Displayed is the low-altitude (850 hPa) component of the Kelvin wave's zonal wind. (**c**) The amplitude of the observed (solid blue line) and model-predicted (dashed red line) Kelvin wave averaged from 1980 to 2009. Reproduced from Stechmann and Ogrosky (2014)

tied to monsoons, the El Niño–Southern oscillation (ENSO), and other observed phenomena in the tropics (Julian and Chervin 1978; Barber and Chavez 1983).

Since the Walker circulation is an overturning of air, there are simple models for this circulation that use only the first baroclinic vertical mode of winds and potential temperature. The fact that such simple models exist and that they are closely tied to the MJO skeleton model makes it a convenient starting point for introducing the data analysis methods that will be used in this chapter. One such model is a steady-state version of the Matsuno–Gill model (Matsuno 1966; Webster 1972; Gill 1980; Gill and Rasmusson 1983; Heckley and Gill 1984), which consists of the first baroclinic linearized shallow-water equations on an equatorial beta plane. We now briefly describe its derivation (additional details may be found in, e.g., Majda 2003; Stechmann and Majda 2015).

5.2.1 Model Derivation

The starting point is the set of linear long-wave equatorial primitive equations,

$$\frac{\partial u}{\partial t} - yv + \frac{\partial p}{\partial x} = f^u, \tag{5.1a}$$

$$yu + \frac{\partial p}{\partial y} = 0, \tag{5.1b}$$

$$\frac{\partial p}{\partial z} = \theta, \tag{5.1c}$$

$$\frac{\partial u}{\partial x} + \frac{\partial v}{\partial y} + \frac{\partial w}{\partial z} = 0, \tag{5.1d}$$

$$\frac{\partial \theta}{\partial t} + w = f^\theta, \tag{5.1e}$$

where (u, v, w) are the zonal, meridional, and vertical velocity anomalies, respectively, and p and θ are the pressure and potential temperature, respectively. The zonal momentum source is f^u and the potential temperature source is f^θ. The time tendency term v_t of meridional wind is neglected due to the assumed equatorial long-wave scaling. The equations have been made dimensionless using standard equatorial reference scales (Stechmann and Majda 2015).

To arrive at a two-dimensional form of (5.1), vertical basis functions can be used to separate (5.1) into an infinite set of shallow-water systems for the baroclinic modes. Here, only the first baroclinic mode is retained so that $u(x, y, z, t) = \tilde{u}(x, y, t)\sqrt{2}\cos(z)$, etc. The resulting first baroclinic equations are (dropping tildes for ease of notation)

$$\frac{\partial u}{\partial t} - yv - \frac{\partial \theta}{\partial x} = 0, \tag{5.2a}$$

$$yu - \frac{\partial \theta}{\partial y} = 0, \tag{5.2b}$$

$$\frac{\partial \theta}{\partial t} - \frac{\partial u}{\partial x} - \frac{\partial v}{\partial y} = f^\theta. \tag{5.2c}$$

where it has been assumed that the momentum source f^u is small compared to the heating source f^θ; this momentum source will be neglected hereafter.

To obtain a one-dimensional form of (5.2), meridional basis functions can be used to separate (5.2) into an infinite set of equations including the Kelvin wave $K(x)$ and long-wave equatorial Rossby waves $R_1(x), R_2(x), R_3(x)$, etc. The natural meridional basis functions are the parabolic cylinder functions:

$$\phi_m(y) = \frac{1}{(m!\sqrt{\pi}\,2^m)^{1/2}} H_m(y) e^{-y^2/2}, \quad m = 0, 1, 2, \ldots, \tag{5.3}$$

where $H_m(y)$ are the Hermite polynomials:

$$H_m(y) = (-1)^m e^{y^2} \frac{d^m}{dy^m} e^{-y^2}. \tag{5.4}$$

For example, explicit formulas for the first few are

$$\phi_0(y) = \frac{1}{\pi^{1/4}} e^{-y^2/2}, \tag{5.5a}$$

$$\phi_1(y) = \frac{1}{\pi^{1/4}} \frac{1}{\sqrt{2}} (2y) e^{-y^2/2}, \tag{5.5b}$$

$$\phi_2(y) = \frac{1}{\pi^{1/4}} \frac{1}{2\sqrt{2}} (4y^2 - 2) e^{-y^2/2}. \tag{5.5c}$$

The functions $\phi_m(y)$ form an orthonormal basis, and the variable $u(x, y)$ can be expanded as

$$u(x, y, t) = \sum_{m=0}^{\infty} u_m(x, t) \phi_m(y), \tag{5.6}$$

where the quantities $u_m(x)$ are obtained using the projection

$$u_m(x, t) = \int_{-\infty}^{\infty} u(x, y, t) \phi_m(y)\, dy. \tag{5.7}$$

Formulas analogous to (5.6) and (5.7) also apply to v, θ, and f^θ. Utilizing the projections in (5.7), the wave amplitudes $K(x, t)$ and $R_m(x, t)$ may be defined as

$$K(x, t) = \frac{1}{\sqrt{2}} (u_0 - \theta_0), \tag{5.8a}$$

$$R_m(x, t) = \frac{\sqrt{m+1}}{\sqrt{2}} (u_{m+1} - \theta_{m+1}) - \frac{\sqrt{m}}{\sqrt{2}} (u_{m-1} + \theta_{m-1}), \quad m = 1, 2, 3, \ldots. \tag{5.8b}$$

Since the Walker circulation is a steady background circulation that is expected to change significantly only on timescales of seasons or more, Eqs. (5.2) may be simplified by omitting the time tendency terms u_t and θ_t. With these terms omitted, and utilizing both the projection onto meridional basis functions and the wave variables K and R_m, (5.2) may be written as

$$\frac{dK}{dx} = -\frac{1}{\sqrt{2}} f_0^\theta,$$ (5.9a)

$$\frac{dR_m}{dx} = (2m+1)v_m - \frac{\sqrt{m+1}}{\sqrt{2}} f_{m+1}^\theta + \frac{\sqrt{m}}{\sqrt{2}} f_{m-1}^\theta, \quad m = 1, 2, 3, \ldots.$$ (5.9b)

Equations (5.9) are a model for the Walker circulation in terms of the heating source term f^θ and meridional winds v_m; note that if a closure equation is provided for v_m, these equations are also steady-state versions of the K and R_m equations in the MJO skeleton model.

We next wish to use this model (5.9) and assess how well it is able to describe the observed Walker circulation. Of course, in order to use the model, the source terms must be prescribed. Likewise, in order to assess how well the model describes the atmosphere's observed response to these source terms, we must estimate the terms K and R_m in observational data. We turn to these tasks next.

5.2.2 Estimating f^θ, K, and R_m

An estimate of f^θ will be constructed using NOAA interpolated outgoing long-wave radiation (OLR) (Liebmann and Smith 1996); interpolated OLR data is provided by the NOAA/OAR/ESRL PSD, Boulder, Colorado, USA, from their website at http://www.esrl.noaa.gov/psd/. To identify observational estimates of the variables K and R_m in (5.9), NCEP/NCAR reanalysis daily winds and geopotential height at two pressure levels (850 and 200 hPa) are used (Kalnay et al. 1996); this data is available from the same website as interpolated OLR. Both datasets have a spatial resolution of $2.5° \times 2.5°$ and the time period used in this study is from 1 January 1979 through 31 December 2011.

To construct our estimate of f^θ, we assume that anomalies in OLR are directly proportional to f^θ, i.e.,

$$f^\theta(x, y, t) = -H_{\text{OLR}} \cdot \text{OLR}(x, y, t).$$ (5.10)

The proportionality constant that provides the best agreement between observations and model results for K is

$$H_{\text{OLR}} = 0.056 \text{ K day}^{-1}(\text{Wm}^{-2})^{-1}.$$ (5.11)

To estimate the meridional modes f_m^θ of this heating source, the data $\text{OLR}(x, y, t)$ is projected onto the meridional modes according to (5.6) and (5.7) (with OLR in place of u). Next, only anomalies from a zonal mean are retained, i.e.,

$$\tilde{f}_m^\theta(x, t) = f_m^\theta - \int_0^{P_E} f_m^\theta \, dx,$$ (5.12)

where P_E is the circumference of the Earth (and dropping tildes hereafter); note that this step is necessary in order to obtain a periodic function for K and R_m in (5.9). Finally, steady versions of each variable are constructed by averaging each $f_m^\theta(x, t)$ over the entire 33-year time period used.

Substitution of f_m^θ into (5.9) and integrating each equation in x results in model predictions of $K(x)$ and $R_m(x)$. Figure 5.1c shows the model prediction of $K(x)$; the winds associated with K may be reconstructed according to

$$u_K(x) = \frac{K(x)}{\sqrt{2}}\phi_0(y), \tag{5.13}$$

and are depicted in Fig. 5.1b.

We next check to see if this accurately describes the Kelvin component of the Walker circulation by estimating K directly from reanalysis data. In order to estimate $K(x, t)$ (as well as $R_m(x, t)$) in observations, we systematically convert reanalysis data $u(x, y, z, t)$, $v(x, y, z, t)$, and $\theta(x, y, z, t)$ data into K and R_m by following the series of projections described above. First, we identify an estimate of first baroclinic winds $u(x, y, t)$, $v(x, y, t)$, and potential temperature $\theta(x, y, t)$ using the simple two-level projection

$$u(x, y, t) = \frac{U850(x, y, t) - U200(x, y, t)}{2\sqrt{2}}, \tag{5.14a}$$

$$v(x, y, t) = \frac{V850(x, y, t) - V200(x, y, t)}{2\sqrt{2}}, \tag{5.14b}$$

$$\theta(x, y, t) = \frac{Z850(x, y, t) - Z200(x, y, t)}{2\sqrt{2}}, \tag{5.14c}$$

where $U850$ refers to zonal winds at 850 hPa, etc., and where Z is geopotential height and is related to potential temperature through a hydrostatic balance equation

$$\frac{\partial Z}{\partial p} = -\theta, \tag{5.15}$$

see, e.g., Stechmann and Majda (2015) for more details.

Next, each variable is decomposed into a sum of its meridional mode components by projection onto the meridional basis functions as described by Eqs. (5.6) and (5.7), resulting in datasets $u_m(x, t)$, $v_m(x, t)$, and $\theta_m(x, t)$. The wave variables $K(x, t)$ and $R_m(x, t)$ are estimated by Eqs. (5.8). Finally, steady versions $K(x)$ and $R_m(x)$ of these variables are estimated by taking the average over the entire 33-year time period.

5.2.3   Results

Figure 5.1c shows the observed Kelvin component of the Walker circulation. The agreement between observations and the model is remarkable. We note that this agreement has been found with *no damping* terms included in the model. This is surprising given that many previous studies have often required large damping terms to find agreement between models and observations; it has been noted by several studies that the need for strong damping seems unsettling (Neelin 1988; Battisti et al. 1999). (The apparent discrepancy between the current results and those of previous studies can be explained in part by the fact that (5.9) is a model for the zonal *anomalies* only, i.e., the $k = 0$ mode of K is not prescribed by the model; more discussion of this point can be found in the supporting materials of Stechmann and Ogrosky (2014).) The agreement for the first two Rossby waves is also remarkable (not shown); this agreement declines with increasing meridional mode number which is perhaps to be expected as higher meridional modes extend further away from the tropics where the assumptions underlying the model begin to break down (Stechmann and Ogrosky 2014).

The agreement between observed and model K is remarkable even over shorter time intervals like an individual season. Figure 5.2c, d shows the model-obs agreement over the June–July–August (JJA) 1997 season and December–January– February (DJF) 1998–1999 seasons, respectively. This agreement is especially noteworthy given that these two seasons are marked by strong El Niño and La Niña events, respectively, corresponding to significant changes in the Walker circulation.

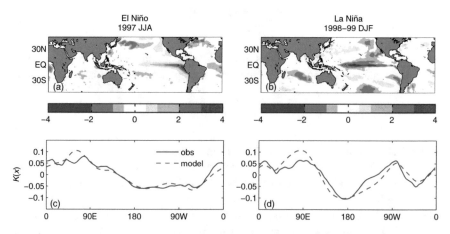

Fig. 5.2 The Kelvin wave amplitude during example 3 month segments with contrasting ENSO state. Sea surface temperature anomalies (K) are shown for (**a**) June–July–August of 1997 during a strong El Niño event and (**b**) December–January–February of 1998–1999 during the subsequent La Niña event. The time-averaged amplitude $K(x)$ of the observed (solid blue line) and model-predicted (dashed red line) Kelvin wave is shown for (**c**) the 1997 JJA El Niño event and (**d**) the 1998–1999 DJF La Niña event. Reproduced from Stechmann and Ogrosky (2014)

5.3 An MJO Index (MJOS) Based on the Skeleton Model

In this section, we extend the data analysis method described above to the skeleton model and see how strongly the theoretical structure of the skeleton model is seen in reanalysis data. We begin by recalling the skeleton model in its simplest form and its primary features.

5.3.1 Solutions to the Linear MJO Skeleton Model

The MJO skeleton model is a nonlinear oscillator model which couples the time-dependent version of the Walker circulation model (5.9) to two additional evolution equations: one for lower-tropospheric moisture Q and one for the planetary-scale wave envelope of convective activity A. In its simplest form, the model is truncated to retain only the zeroth meridional mode of A; this model contains exactly one nonlinear term proportional to QA. If a constant background state of convective activity $A_s = \bar{A}$ is removed from the model, the remaining terms which describe anomalies from this background state may be linearized around A_s, resulting in the coupled system of linear PDEs,

$$\frac{\partial K}{\partial t} + \frac{\partial K}{\partial x} = -\frac{\bar{H}}{\sqrt{2}} A, \tag{5.16a}$$

$$\frac{\partial R}{\partial t} - \frac{1}{3}\frac{\partial R}{\partial x} = -\frac{2\sqrt{2}\bar{H}}{3} A, \tag{5.16b}$$

$$\frac{\partial Q}{\partial t} + \frac{\tilde{Q}}{\sqrt{2}}\frac{\partial K}{\partial x} - \frac{\tilde{Q}}{6\sqrt{2}}\frac{\partial R}{\partial x} = \left(\frac{\tilde{Q}}{6} - 1\right)\bar{H}A, \tag{5.16c}$$

$$\frac{\partial A}{\partial t} = \Gamma Q A_s, \tag{5.16d}$$

where Q and A are zeroth meridional mode components of the moisture and wave envelope, respectively, R is the first Rossby wave (denoted R_1 in the previous section), $\tilde{Q}$ is a background vertical moisture gradient, Γ is a growth/decay rate for the wave envelope of convective activity, and $\bar{H}$ is a heating rate prefactor that is irrelevant to the dynamics.

This linear system is readily solved by expressing each variable as a superposition of plane waves, e.g.,

$$K(x,t) = \sum_{k=-\infty}^{\infty} \hat{K}(k)\exp\left(\frac{2\pi i k x}{P_E} - i\omega t\right), \tag{5.17}$$

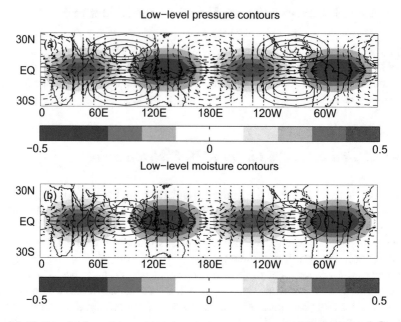

Fig. 5.3 Uniform background state MJO-2 mode, i.e., $\alpha = 0$; $\omega = 0.0235$ cycles day^{-1}, $\bar{k} = 2.0$. Convective anomalies are depicted by shading. Solid (dashed) lines denote positive (negative) (**a**) pressure anomalies and (**b**) moisture anomalies. Reproduced from Ogrosky and Stechmann (2015a)

with similar expressions for R, Q, and A, where ω is a temporal frequency; the system may then be solved for each zonal wavenumber k. For every k, there are four solutions each consisting of an eigenvalue ω and eigenvector $\hat{\mathbf{e}} = (\hat{e}_K, \hat{e}_R, \hat{e}_Q, \hat{e}_A)^T$. Recall that one of these four solutions has structure and propagation reminiscent of the observed MJO (Majda and Stechmann 2009). Figure 5.3 shows the solution for the $k = 2$ MJO mode with $\omega = 0.0235$ cycles day^{-1}, corresponding to a phase speed of $5.4 \, \mathrm{m \, s}^{-1}$.

Several questions might now be asked about the model, including: "How clearly is the pattern of winds, pressure, moisture, and convective activity shown in Fig. 5.3 seen during strong observed MJO events? Is this pattern also seen during times when the MJO is not active? If this pattern is strongly correlated with observed MJO activity, can we use the theory to identify MJO events in nature?" The rest of this section will be devoted to addressing these questions using an extended form of the data analysis techniques discussed in the previous section.

5.3.2 Data Analysis Methods

We have already seen how to identify K and R_m in reanalysis data for a steady-state model of the Walker circulation; likewise we have used OLR to identify A. The same approaches may be used to identify K, R, and A in the skeleton model, but we no longer average the data over a time period. Instead, each day's data is used to create a daily snapshot of each variable so that each variable is a function of both longitude and time, i.e., $K(x, t)$, $R(x, t)$, and $A(x, t)$.

We next need an estimate of $Q(x, t)$. Note that Q represents lower-tropospheric moisture and is *not* a first baroclinic variable. Accordingly, we use an average of specific humidity data across several pressure levels near the bottom of the troposphere:

$$Q_{LT}(x, y, t) = \frac{1}{4} Q_{925\text{hPa}} + \frac{1}{2} Q_{850\text{hPa}} + \frac{1}{4} Q_{725\text{hPa}}. \tag{5.18}$$

Next, Q_{LT} is decomposed into a sum of its meridional mode components by projection onto the meridional basis functions as described by Eqs. (5.6) and (5.7), (with Q replacing u), resulting in datasets $Q_m(x, t)$.

We would next like to know how prominently the skeleton model's MJO structure appears in reanalysis data. To assess this, the zonal Fourier modes $\hat{\mathbf{U}}(k) = (\hat{K}(k), \hat{R}(k), \hat{Q}(k), \hat{A}(k))^T$ of each day's $\mathbf{U} = (K, R, Q, A)$ data are projected onto the model's corresponding MJO eigenvector, i.e.,

$$\widehat{MJOS}(k) = \hat{\mathbf{e}}_{MJO}(k)^\dagger \hat{\mathbf{U}}(k). \tag{5.19}$$

where $\mathbf{W}^\dagger$ denotes the conjugate transpose of $\mathbf{W}$, and where the time dependence has been suppressed to ease notation. The inverse Fourier transform then leads to the real-valued scalar quantity

$$\text{MJOS}(x, t) = \text{Projection of } (K(x, t), R(x, t), Q(x, t), A(x, t))^T$$

$$\text{onto the MJO skeleton eigenvector,} \tag{5.20}$$

which we refer to as the MJO skeleton signal. Finally, to succinctly quantify the strength of $\text{MJOS}(x, t)$ variability, its amplitude $\text{MJOSA}(t)$ may be computed as a root-mean-square of the zonal variations,

$$\text{MJOSA}(t) = \left[\frac{1}{P_E} \int_0^{P_E} |\text{MJOS}(x, t)^2| \, dx \right]^{1/2}. \tag{5.21}$$

5.3.3   Results

Three case studies are now considered to illustrate the range of variability of $\text{MJOS}(x, t)$. One case is the well-known period of strong MJO activity from 1 July 1987 to 1 July 1988 (Hendon and Liebmann 1994; Wheeler and Hendon 2004). This is the middle case shown in Fig. 5.4. The other two cases of Fig. 5.4 are periods

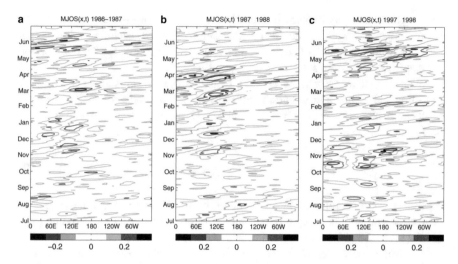

Fig. 5.4 MJO skeleton signal, MJOS(x, t), for the 1 year periods (**a**) 1 July 1986 through 30 June 1987, (**b**) 1 July 1987 through 30 June 1988, (**c**) 1 July 1997 through 30 June 1998. Units are nondimensional. Corresponding plots of amplitude MJOSA(t) are shown in Fig. 5.5. Reproduced from Stechmann and Majda (2015)

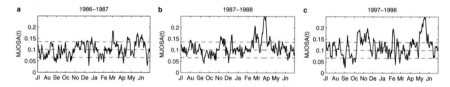

Fig. 5.5 Amplitude MJOSA(t) of the MJO skeleton signal. The 1-year periods from (**a**) 1 July 1986 through 30 June 1987, (**b**) 1 July 1987 through 30 June 1988, (**c**) 1 July 1997 through 30 June 1998 are shown. Units are nondimensional. The three dashed lines indicate the climatological mean (middle) and the mean plus/minus one standard deviation (top/bottom). Corresponding MJOS(x, t) signal is shown in Fig. 5.4. Reproduced from Stechmann and Majda (2015)

when the presence and/or initiation of MJO activity is somewhat ambiguous (Straub 2013). On the left is the period from 1 July 1986 to 1 July 1987, and on the right is the period from 1 July 1997 to 1 July 1998, which includes the time of May–June 1998 of the South China Sea Monsoon Experiment (SCSMEX). The corresponding MJOSA(t) signal is shown in Fig. 5.5 for each time period.

In the 1987–1988 case, several periods show strong variability. The MJOS(x, t) signal is strongest in November–December and February–March, with a moderate signal in the interim during January. An interesting feature is the circumnavigating signal through February, March, and April. The western hemisphere component of the signal could be a strong instance of the fast "dry" Kelvin wave signals described by Milliff and Madden (1996), although in this case one can discern concomitant signals in all of the four variables K, R, Q, and A (not shown), which suggests this is an instance of coupled convection-circulation variability.

The 1986–1987 case has been ambiguous in earlier work. Wavenumber–frequency-filtered OLR indicates a strong active MJO event throughout January and February, whereas the real-time multivariate MJO (RMM) index (Wheeler and Hendon 2004) indicates a strong MJO event from late February to early March (Straub 2013). Here, MJOS(x, t) does not indicate a particularly strong signal for either of those two periods. A short burst appears at the beginning of March, although there is no particularly strong MJO event indicated by MJOS(x, t) during January, February, and March. It seems that the K, R, Q, and A signals are not sufficiently coupled to indicate a coupled convection-circulation structure at these times.

The final case of 1997–1998 has also been ambiguous in earlier work with other MJO indices (Straub 2013). Complicating the picture is a synoptic-scale convectively coupled Kelvin wave that appears in mid-May (Straub et al. 2006; Straub 2013); its rapidly propagating signature is evident in MJOS(x, t) in Fig. 5.4c near the dateline, from 150°E to 150°W. However, in late April and early May, a planetary-scale signal is also present in MJOS(x, t). Its strongest anomalies are negative from the dateline to 60°W and positive from 0°E to 90°E; and its slower propagation speed is consistent with the MJO. From this information in the MJOS(x, t) signal, it is possible that an MJO is present in early May and then transforms into the convectively coupled Kelvin wave in mid-May; or it is possible that the MJO signal is present throughout May but is contorted by the concomitant signal of the convectively coupled Kelvin wave. The presence of an MJO is also suggested by a strong Rossby signal $R(x, t)$ that is present during May and propagates slowly eastward (not shown).

An interesting feature in Fig. 5.4 is the mixture of propagating and standing signals. For example, two periods of standing signals are December 1986 in Fig. 5.4a and October 1997 in Fig. 5.4c. This variety suggests that MJOS(x, t) includes a range of intraseasonal variability types. It should also be noted that MJOS(x, t) variability can be strong in any season. This suggests that the structures of intraseasonal variability—in both boreal summer and winter—have a significant projection onto the theoretical coupled convection-circulation structures used here.

5.4   A Warm-Pool MJO Index (MJOS$_k$) Based on the Skeleton Model

In the previous section, the skeleton model was linearized about a constant background state of convection, A_s; i.e., $A_s = \bar{A}$ was assumed to be a constant rather than a function of x. Observations and reanalysis data suggest, however, that A_s may vary significantly with longitude. The goal of this section is to once again find the eigenvectors of the linear skeleton model, but this time we will use an observation-based estimate of the background state and allow for zonal variations in this parameter, i.e., $A_s = A_s(x)$.

5.4.1   Estimating $A_s(x)$

The background state $A_s(x)$ may be estimated by averaging the quantity $A(x, t)$ over a sufficiently long time period, i.e.,

$$A_s(x) = \frac{1}{T} \int_0^T A(x, t) \, dt. \tag{5.22}$$

In the previous section, A was estimated using OLR. This is not, however, the only quantity that can be used to estimate A, and we take the opportunity here to consider another choice, namely precipitation. (There are good reasons for using precipitation data rather than OLR, but a discussion of this choice lies somewhat outside the scope of the current text, and the differences in the results discussed here created by using precipitation rather than OLR are not large.)

Global Precipitation Climatology Project (GPCP) daily precipitation data, recorded in mm day^{-1} with spatial resolution of $1° \times 1°$ from 1 January 1997 to 31 December 2013 (Huffman et al. 2012), is used to estimate convective heating $A_s(x)$ as follows. The energy released by M mm day^{-1} of precipitation at a given location increases the temperature of the surrounding column of air at a rate of

$$\bar{H} A = \left(\frac{g \rho_w L_v}{p_0 c_p} \right) M, \tag{5.23}$$

where $p_0 = 1.013 \times 10^5$ kg m^{-1} s^{-2} is the mean atmospheric pressure at mean sea level, $\rho_w = 10^3$ kg m^{-3} is the density of water, $g = 9.8$ m s^{-2} is acceleration due to gravity, and (5.23) has units of K day^{-1}. An average precipitation of 1 mm day^{-1} thus corresponds to a heating rate of ≈ 0.24 K day^{-1} or a dimensionless heating rate of $\bar{H} A \approx 0.0054$ when scaled by the characteristic heating rate of 45 K day^{-1}. This 2D (x, y) heating rate is converted to 1D (x) by projecting onto the meridional modes in the same manner as the primitive variables u, θ, etc.; A is then estimated by the leading meridional mode, i.e.,

$$\bar{H} A(x, t) = \bar{H} A_0(x, t). \tag{5.24}$$

Averaging over the 17-year time period (either annually or seasonally) results in the background states $A_s(x)$ shown in Fig. 5.6. In practice, zonal variations are reduced in amplitude so that the background state is more nearly uniform, i.e.,

$$A_s(x) = \hat{S}_0 + \sum_{k=1}^N \alpha \left[\hat{S}_k \exp \left(\frac{2\pi i k x}{P_E} \right) + \hat{S}_{-k} \exp \left(\frac{-2\pi i k x}{P_E} \right) \right], \tag{5.25}$$

where $\alpha \leq 1$. This change is made as it turns out that the eigenvectors to be considered are quite sensitive to these zonal variations; see Ogrosky and Stechmann (2015a) for further discussion of this. In addition, the background state is smoothed

Fig. 5.6 Time-averaged background convective state $\bar{H}A_s(x)$ calculated from GPCP data. Data averaged from 1 January 1997 to 31 December 2013; annual, DJF, and JJA data shown in black, red, and blue, respectively. All quantities are dimensionless. Reproduced from Ogrosky and Stechmann (2015a)

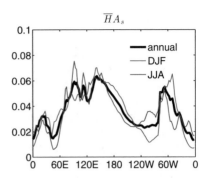

to retain only the first N Fourier modes, so that $\hat{S}_j = 0$ for $|j| > N$; in all cases presented below, $N = 1$.

Substitution of (5.25) into (5.16) yields a system of $8k_m + 4$ equations due to the four variables that are each expanded in terms of $2k_m + 1$ wavenumbers k with $-k_m \leq k \leq k_m$. This is a single system of equations where each wavenumber is coupled to every other wavenumber through the background state $A_s(x)$. The case of a constant background state studied by Majda and Stechmann (2009) is recovered if we take $N = 0$, i.e., if $\hat{S}_j = 0$ for $j \neq 0$. In this case a system of 4 equations for wavenumber k_{i_1} decouples from the system for k_{i_2}, with $i_1 \neq i_2$, so that the problem becomes $2k_m + 1$ eigenvalue problems, each for a different wavenumber; each of these can then be solved separately, and the resulting eigenvectors each contain contributions from one wavenumber only. Here, in contrast, $N > 0$ and each system of four equations is coupled to the other systems through the background state A_s. Additional details and further discussion of the problem and solution procedure can be found in Ogrosky and Stechmann (2015a). We note that other studies have examined the skeleton model in the presence of a sinusoidally varying background state (e.g., Majda and Stechmann 2011; Thual et al. 2014).

5.4.2   Model Solutions

The solutions to (5.16) with zonally varying $A_s(x)$ are once again linear modes of 4 types: dry Kelvin modes, dry Rossby modes, moist Rossby modes, and MJO modes. Each MJO mode has an averaged-wavenumber, $\bar{k}$, which can be calculated by

$$\bar{k} = \sum_{k=-k_m}^{k_m} |k|\sqrt{\hat{K}_k^2 + \hat{R}_k^2 + \hat{Q}_k^2 + \hat{A}_k^2}. \tag{5.26}$$

In the case with uniform $A_s = \overline{A_s}$, $\bar{k}$ is an integer for each mode; in the case with zonally varying $A_s(x)$, $\bar{k}$ is a real number that assesses on average what wavenumber contains the most power. The MJO modes can then be ordered by average wavenumber, i.e., they will be denoted by

$$\hat{\mathbf{e}}_1, \hat{\mathbf{e}}_2, \ldots, \tag{5.27}$$

so that $\hat{\mathbf{e}}_1$ refers to the MJO mode with smallest $\bar{k}$, $\hat{\mathbf{e}}_2$ refers to the MJO mode with $\bar{k}$ larger than that of $\hat{\mathbf{e}}_1$ and smaller than $\bar{k}$ for all other MJO modes, and so on. In some places in the text, the alternate notation MJO-1, MJO-2, etc., may be used. As the skeleton model was proposed to describe the planetary-scale dynamics of the MJO, it is the lowest-wavenumber modes, i.e., $\hat{\mathbf{e}}_j$ for say $j = 1$–4, that are of the most relevance here; only these first few modes with small $\bar{k}$ will be studied here.

Many of the features seen in the uniform modes are reproduced in the varying modes, e.g., the convective activity A is in quadrature with the other three variables K, R, and Q, and A lags behind Q, R, and a pair of anticyclones, and leads K and a pair of cyclones. However there is additional structure created by the coupling between wavenumbers; both the amplitudes and phases of the components of different wavenumbers are fixed relative to each other. This additional structure creates a wave envelope with maximum amplitude centered over the maximum of the background convective heating anomaly located at approximately 140°E. This wave envelope can be seen in Fig. 5.7 which shows four snapshots of the convective activity of the MJO-2 mode; the envelope has a base of support stretching from roughly 60°E to 160°W, spanning the Indian Ocean, Maritime Continent, and western Pacific Ocean. Linear disturbances propagate eastward through this envelope with phase speed $\approx 4 \text{ m s}^{-1}$, increasing in strength over the Indian Ocean, reaching maximum amplitude near the western Pacific warm pool, and decaying

Fig. 5.7 Propagation of varying MJO-2 mode convective anomalies (shading) with $\alpha = 0.1$, $N = 1$; (**a**) $t = 0$ days, (**b**) $t = 10$ days, (**c**) $t = 20$ days, (**d**) $t = 30$ days. Reproduced from Ogrosky and Stechmann (2015a)

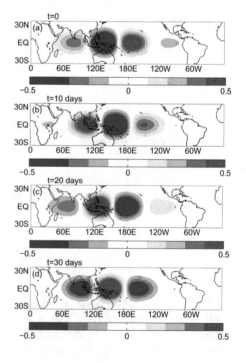

rapidly after crossing the dateline. In contrast, the sinusoidal uniform modes have anomalies over the global tropics.

5.4.3 Eigenvector Projection

In the previous section, the uniform MJO modes were used to identify the MJO in reanalysis data. This was achieved in two steps: first, reanalysis data $\mathbf{U}(x, t) = (K, R, Q, A)^{\mathrm{T}}(x, t)$ was projected onto each of the low-wavenumber MJO modes, i.e., (5.19) for $k = 1$–3. Thus for each wavenumber k we have a measure of the strength and phase of the MJO-k signal. Second, taking the inverse Fourier transform of (5.19) gives the real-valued scalar quantity MJOS(x, t), referred to as the MJO skeleton signal. The reanalysis data was found to contain a strong signal of this theoretical structure at times and locations where well-documented observed MJOs have occurred over the last 30 years.

The first step described above, i.e., projecting reanalysis data onto individual MJO modes, may be extended to the varying background case with little modification. We now consider reanalysis data

$$\hat{\mathbf{U}} = [\hat{\mathbf{U}}_{-k_m}, \ \ \hat{\mathbf{U}}_{-k_m+1}, \ \ \ldots \ \ \hat{\mathbf{U}}_{k_m-1}, \ \ \hat{\mathbf{U}}_{k_m}]^{\mathrm{T}}, \tag{5.28}$$

with

$$\hat{\mathbf{U}}_k = [\hat{K}_k, \ \ \hat{R}_k, \ \ \hat{Q}_k, \ \ \hat{A}_k]^{\mathrm{T}}. \tag{5.29}$$

This data is projected onto a single MJO eigenvector $\tilde{\mathbf{e}}_j$ in a manner similar to (5.19),

$$\begin{aligned} \mathrm{MJOS}_j(t) &= \tilde{\mathbf{e}}_j^{\dagger} \hat{\mathbf{U}}(t) \\ &= \mathrm{MJOSA}_j(t) \exp[i\,\mathrm{MJOSP}_j(t)], \end{aligned} \tag{5.30}$$

resulting in a complex-valued scalar function of time, where the magnitude of this signal is denoted by MJOSA$_j(t)$ and the phase of the signal is denoted by MJOSP$_j(t)$. (Note that here we use MJOS, rather than $\widetilde{MJOS}$ as in Stechmann and Majda (2015), to denote the signal in Fourier space.) The results in this section were found using the MJO modes with $\alpha = 0.1$ and $N = 1$, and using reanalysis data smoothed to retain only wavenumbers $k = \pm 1, \pm 2, \pm 3$. In all figures, the signal MJOS$_j$ has been normalized by its standard deviation over the 16 year period 1 July 1997 through 30 June 2013; an MJOSA value of one thus indicates a signal one standard deviation away from the complex mean of approximately zero.

The evolution of MJOS$_j$, MJOSA$_j$, and accumulated MJOSP$_j$ is shown in Fig. 5.8 for $j = 2$ from 1 July 2009 through 30 June 2010. This period overlaps with the Year of Tropical Convection (YOTC) (Moncrieff et al. 2012; Waliser et al. 2012). The most significant MJO-like activity during YOTC occurred during the

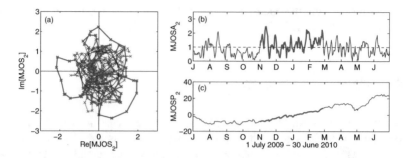

Fig. 5.8 (**a**) Evolution of the complex-valued $MJOS_j(t)$ for $j = 2$ from 1 July 2009 to 30 June 2010. Here $MJOS_j$ has been rescaled by its standard deviation; points outside the circle are greater than 1 standard deviation away from the mean. (**b**) Evolution of $MJOSA_j$ and (**c**) $MJOSP_j$ for $j = 2$ over the same time period. Thick red lines denote the period from 1 November 2009 to 26 February 2010. Reproduced from Ogrosky and Stechmann (2015a)

2009–2010 DJF season. The $MJOSA_2$ signal in panel (b) is strongest in early November and late January through February of this season. The variations in signal strength from late November through early January indicate that the observations only partially reflect the theoretical structure of the MJO skeleton model with varying background state during this time. In panel (c), an increase of 2π in the accumulated phase indicates the signal MJOS has completed one counterclockwise cycle around the complex plane. Periods of elevated MJOSA signal correspond with an upward trend in the accumulated phase $MJOSP_2$; the increase indicates eastward propagation. Note that the accumulated phase is only meaningful during time periods of a strong MJOSA signal. In panel (a), these two trends are shown together by plotting the daily values of the complex-valued signal $MJOS_2$.

The pattern correlation can be used to compare the $MJOSA_2$ signal with other MJO indices currently in use and is here defined as

$$PC(f, g) = \frac{\frac{1}{T} \int_0^T f(t)g(t)dt}{\left[\frac{1}{T} \int_0^T |f(t)|^2 dt\right]^{1/2} \left[\frac{1}{T} \int_0^T |g(t)|^2 dt\right]^{1/2}}, \tag{5.31}$$

where T is the length of time considered and f and g are two MJO indices defined on the time period $[0, T]$. During the time period shown in Fig. 5.8, i.e., 1 July 2009 through 30 June 2010, the $MJOSA_2$ signal has a pattern correlation of 0.86 with the RMM index (Wheeler and Hendon 2004) and a pattern correlation of 0.87 with the OLR MJO index (OMI) (Kiladis et al. 2014); the RMM and OMI have a pattern correlation of 0.90 with each other during the same time period. Note that in contrast to the RMM and OMI indices, the $MJOSA_2$ does not rely on empirical orthogonal functions (EOFs) and does not use temporal filtering.

Before concluding this section, we briefly consider how well the stochastic form of the skeleton model reproduces observed variability in the model. Recall that the stochastic model contains a birth/death process in the A equation which is included to represent the effects of unresolved smaller-scale processes on the planetary scale evolution (Thual et al. 2014). As discussed earlier, this stochasticity results in model solutions having realistic event intermittency, including the existence of both wave trains consisting of successive MJO events and extended periods of dormancy. As an example of this intermittency and its agreement with observations, Fig. 5.9 shows a Hovmoller plot of filtered convective activity A in the model and the zeroth meridional mode of observed precipitation over seven boreal winters. The intermittency in the model solutions is in qualitatively good agreement with intermittency seen in nature. (Note that the model and observations should not be directly compared at individual times.) While propagation can be seen in both the eastward and westward directions, the eastward propagation appears to be the dominant feature in both observed and modeled $\bar{H} A$. In the model data, the strongest anomalies appear between day 500 and 720 as two somewhat distinct periods of MJO activity. Individual anomalies have similar strength and propagation speed to the observed anomalies. The intermittency in MJO activity seen in the model is also reminiscent of that in the observations, with several successive MJO events followed by periods of smaller anomalies. The largest anomalies in the observed data generally occur from roughly 60°E through 150°W; the strongest anomalies in the model solution occur in a similar range, though strong anomalies can also be seen near South America and the Atlantic Ocean, e.g., day 760 through 840. Note again that the observed and modeled convective activity should be compared statistically, rather than directly day by day.

The daily values of $MJOSA_2$ are displayed in a histogram in Fig. 5.10 for both observed and modeled data. The distribution of the modeled and observed signals is very similar including long tails that correspond to large-amplitude MJO events. We note that the distribution of signal amplitudes when the stochastic model is run with a uniform background state (not shown) is also similar to observations.

5.5   Assessing the Validity of the Equatorial Long-Wave Approximation

Before concluding this chapter, we consider another application of the data analysis methods discussed here. Both the Walker circulation model and the MJO skeleton model are derived in part using the equatorial long-wave approximation; the absence of the $\partial v / \partial t$ term in Eq. (5.2b) is due to this approximation. Equatorial long-wave dynamics are marked by several related characteristics, including small meridional winds, geostrophic balance in the meridional direction, and inertio-gravity waves of small amplitude. Theories which exploit an assumed small ratio of meridional to zonal lengthscales suggest these dynamics are valid on "long" zonal and temporal scales.

One might ask how "long" the zonal and temporal scales must be in order for the equatorial long-wave approximation, and models (like the MJO skeleton

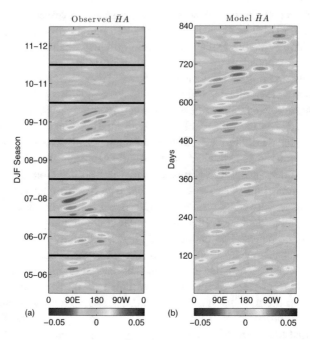

Fig. 5.9 (**a**) Observed and (**b**) modeled $\bar{H}A$ filtered by removing the statistical steady-state and retaining only wavenumbers $k = \pm 1, \pm 2, \pm 3$ and frequencies $1/90 \le \omega \le 1/30$ cpd. Reproduced from Ogrosky and Stechmann (2015a)

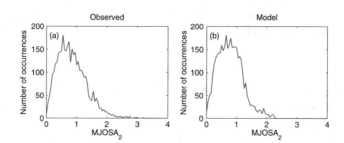

Fig. 5.10 Histogram of $\mathrm{MJOSA}_2(t)$ values for (**a**) observed data from 1 July 2003 to 30 June 2013, and (**b**) model data over an identical length time period (approximately 3650 days). The bin width is 0.05, and 80 bins were used spanning the range of 0–4. Reproduced from Ogrosky and Stechmann (2015a)

model) based on this approximation, to be valid. We close this section by briefly considering this question from two perspectives using the data analysis methods discussed above.

5.5.1 Meridional Winds

In order to determine the scales on which meridional winds are smaller than zonal winds, we will examine the spectral power contained in the anomalies of $u_m(x, t)$, $v_m(x, t)$, and $\theta_m(x, t)$ for various meridional modes m. This standard procedure consists of taking the Fourier transform of each variable in both space and time; this results in a series of Fourier coefficients $\hat{u}_m(k, \omega)$, $\hat{v}_m(k, \omega)$, and $\hat{\theta}_m(k, \omega)$, where k is the zonal wavenumber and ω is the temporal frequency. In practice, additional steps are taken to smooth the spectrum and to account for the fact that the dataset is not periodic in time; see, e.g., Ogrosky and Stechmann (2015b) for details of these additional steps.

Figure 5.11 shows the ratios $|\hat{v}_m(k, \omega)|/|\hat{u}_m(k, \omega)|$ and $|\hat{v}_m(k, \omega)|/|\hat{\theta}_m(k, \omega)|$ for $m = 0$ and 1. Consistent with the long-wave approximation, these ratios are smallest for wavenumber–frequency pairs (k, ω) where both k and ω are small, though the ratio is much smaller than 1 for only a very narrow range of wavenumbers. For example, the solid black contour in panels (a,c) of Fig. 5.11 indicates where $|\hat{v}_m|/|\hat{u}_m| = 0.3$, and the solid black contour in panels (b,d) indicates where $|\hat{v}_m|/|\hat{\theta}_m| = 0.3$. This contour lies within the range $|k| \le 2$ for panel (a) and within the range $|k| \le 1$ for panels (b)-(d).

The dashed line in each panel represents a theoretical prediction of the wavenumber–frequency pairs at which one might expect these ratios to be 0.3 (Ogrosky and Stechmann 2015b). The dark line depicting the 0.3 contour lies entirely within the box for panels (b,c,d), while a portion lies outside the box in panel (a) along a Kelvin wave-type dispersion curve. This theoretical prediction thus appears in general to slightly overestimate the region where the long-wave approximation holds, but the agreement is reasonable.

5.5.2 Meridional Geostrophic Balance

These results suggest that one aspect of the long-wave approximation, small meridional winds, holds only over a very narrow range of scales. We next examine the scales on which meridional geostrophic balance occurs, i.e., the scales on which Eq. (5.2b) holds. This is done most easily by using characteristic variables $r_m(x, t) = (u_m - \theta_m)/\sqrt{2}$ and $l_m(x, t) = (u + \theta_m)/\sqrt{2}$. In terms of r_m and l_m, Eq. (5.2b) is

$$\sqrt{m+1}r_{m+1} + \sqrt{m}l_{m-1} = 0. \tag{5.32}$$

An atmosphere in perfect meridional geostrophic balance will exactly satisfy (5.32) for every m, but in general one may expect (5.32) to only be satisfied approximately. We can assess how well Eq. (5.32) is satisfied by defining the meridional geostrophic *imbalance* as the difference between the right- and left-hand sides of (5.32),

$$MGI_m(x, t) = r_{m+1} + \frac{\sqrt{m}}{\sqrt{m+1}}l_{m-1}. \tag{5.33}$$

Taking the Fourier transform of MGI_m in both x and t allows for a measure of the spectral power $\widehat{MGI}_m(k, \omega)$ at various zonal wavenumbers k and frequencies ω.

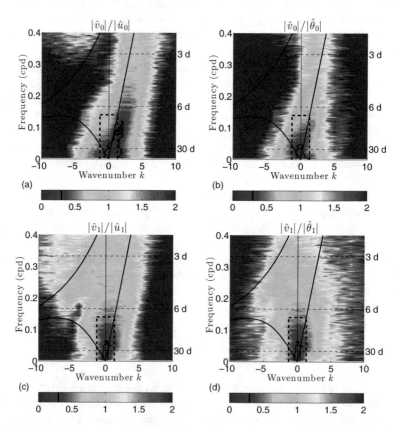

Fig. 5.11 (**a**), (**c**) Power spectrum ratios $|\hat{v}_m(k, \omega)|/|\hat{u}_m(k, \omega)|$ for $m = 0$ and 1, respectively. (**b**), (**d**) Power spectrum ratios $|\hat{v}_m(k, \omega)|/|\hat{\theta}_m(k, \omega)|$ for $m = 0$ and 1, respectively. The dashed black rectangle outlines a theoretical prediction of where one might expect this ratio to be 0.3; the solid black curve denotes the 0.3 contour. Reproduced from Ogrosky and Stechmann (2015b)

An average of MGI_m over low frequencies is shown in Fig. 5.12a, where the quantity

$$\log \left(\frac{1}{\tilde{\omega}} \int_{\omega=0}^{\omega=\tilde{\omega}} |\widehat{MGI}_m(k,\omega)|^2 \right)^{1/2} \tag{5.34}$$

is plotted as a function of wavenumber k, and where an upper frequency cutoff of $\tilde{\omega}/T_E = 0.25$ cpd has been used. This low-frequency power is shown for $m = 1\text{–}6$. For small m, say $m \leq 3$, the power contained in wavenumbers $|k| \leq 3$ is dwarfed by the power in $4 \leq |k| \leq 8$. This trough centered at $k = 0$ becomes less pronounced as m increases.

Note that for $|k| \geq 10$, the low-frequency power of MGI_m is smaller than for $|k| \leq 3$; one might conclude that meridional geostrophic balance is better observed at small scales. However, a better measure for the degree of imbalance is perhaps the ratio of the low-frequency power of MGI_m to the low-frequency power of r_m, i.e., the relative meridional geostrophic imbalance (RMGI),

$$RMGI_m(k) = \left(\frac{1}{\tilde{\omega}} \int_{\omega=0}^{\omega=\tilde{\omega}} |\widehat{MGI}_m(k,\omega)|^2 \right)^{1/2} \bigg/ \left(\frac{1}{\tilde{\omega}} \int_{\omega=0}^{\omega=\tilde{\omega}} |\hat{r}_m(k,\omega)|^2 \right)^{1/2} . \tag{5.35}$$

This quantity is shown in Fig. 5.12b, where (5.35) is plotted as a function of wavenumber k. For all m, this relative imbalance increases approximately monotonically with increasing $|k|$, with a sharp increase beginning at $|k| \approx 4$.

These results suggest that the second aspect of long-wave dynamics considered here (meridional geostrophic balance) is seen in the data over a slightly larger range of spatiotemporal scales than the first aspect (weak meridional winds). Thus, the degree to which a model based on long-wave asymptotics can hope to accurately

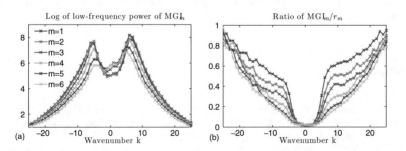

Fig. 5.12 (a) The log of the average low-frequency power of meridional geostrophic imbalance (MGI) as defined in (5.34) with $\tilde{\omega}/T_E = 0.25$ cpd for $m = 1\text{–}6$. (b) The ratio of the average low-frequency power in MGI_m and r_m as defined in (5.35) with $\tilde{\omega}/T_E = 0.25$ cpd for $m = 1\text{–}6$. Reproduced from Ogrosky and Stechmann (2015b)

describe an observed phenomenon may depend on whether it models primitive variables (e.g., u, θ, etc.) or characteristic variables (e.g., r, l, etc.).

5.6   Conclusion

To conclude, the data analysis techniques discussed here have allowed for a sensible comparison between models of tropical atmospheric phenomena like the Walker circulation and the MJO and observational/reanalysis data. Specifically, we have shown that model (5.9), which does not include damping, is an effective model for zonal variations in the components of the Walker circulation, and that OLR is a remarkably good indicator of total diabatic heating on planetary scales. Likewise, solutions to the MJO skeleton model with various forcing functions compare favorably with reanalysis data in several ways.

These model solutions were used to design new ways of analyzing observational data, including the construction of new indices $\text{MJOS}(x, t)$ and $\text{MJOS}_2(t)$. These indices provide a quantitative measure of how well the skeleton theory describes the structure of individual MJO events. These data analysis techniques do not rely on temporal filtering or EOFs, and utilize the eigenvectors of the theory rather than the eigenvalues.

The techniques have also allowed for an assessment of the scales on which long-wave dynamics can be expected to exist in the tropical atmosphere. We note that these methods have other applications as well that lie outside the scope of this text; for example, they have been used to identify some types of convectively coupled equatorial waves (Ogrosky and Stechmann 2016).

The results discussed here, along with many other results in the literature, suggest that data analysis techniques that are grounded in PDE theories for the tropical atmosphere can be powerful tools in the analysis of large observational datasets and the identification and classification of tropical atmospheric phenomena. It is our hope that the discussion and results presented here spur additional work in this direction.

References

Barber RT, Chavez FP (1983) Biological consequences of El Niño. Science 222:1203–1210

Battisti DS, Sarachik ES, Hirst AC (1999) A consistent model for the large-scale steady surface atmospheric circulation in the Tropics. J Clim 12:2956–2964

Gill AE (1980) Some simple solutions for heat-induced tropical circulation. Q J Roy Meteorol Soc 106:447–462

Gill AE, Rasmusson EM (1983) The 1982–83 climate anomaly in the equatorial Pacific. Nature 306:229–234

Heckley WA, Gill AE (1984) Some simple analytical solutions to the problem of forced equatorial long waves. Q J Roy Meteorol Soc 110:203–217

Hendon HH, Liebmann B (1994) Organization of convection within the Madden-Julian oscillation. J. Geophys. Res. 99:8073–8084

Hendon HH, Salby ML (1994) The life cycle of the Madden-Julian oscillation. J Atmos Sci 51:2225–2237

Huffman GJ, Bolvin DT, Adler RF (2012) GPCP Version 2.2 SG Combined Precipitation Data Set. WDC-A, NCDC, Asheville, NC. Data set accessed 12 February 2014 at http://www.ncdc.noaa.gov/oa/wmo/wdcamet-ncdc.html

Julian PR, Chervin RM (1978) A study of the southern oscillation and walker circulation phenomenon. Mon Weather Rev 106:1433–1451

Kalnay E, Kanamitsu M, Kistler R, Collins W, Deaven D, Gandin L, Iredell M, Saha S, White G, Woollen J, Zhu Y, Chelliah M, Ebisuzaki W, Higgins W, Janowiak J, Mo KC, Ropelewski C, Wang J, Leetmaa A, Reynolds R, Jenne R, Joseph D (1996) The NCEP/NCAR 40-year reanalysis project. Bull Am Meteor Soc 77:437–471

Kiladis GN, Dias J, Straub KH, Wheeler MC, Tulich SN, Kikuchi K, Weickmann KM, Ventrice MJ (2014) A comparison of OLR and circulation-based indices for tracking the MJO. Mon Weather Rev 142:1697–1715

Liebmann B, Smith CA (1996) Description of a complete (interpolated) outgoing long wave radiation dataset. Bull Am Meteorol Soc 77:1275–1277

Majda AJ (2003) Introduction to PDEs and waves for the atmosphere and ocean. Courant lecture notes in mathematics, vol 9. American Mathematical Society, Providence

Majda AJ, Stechmann SN (2009) The skeleton of tropical intraseasonal oscillations. Proc Natl Acad Sci 106:8417–8422

Majda AJ, Stechmann SN (2011) Nonlinear dynamics and regional variations in the MJO skeleton. J Atmos Sci 68:3053–3071

Matsuno T (1966) Quasi-geostrophic motions in the equatorial area. J Meteorol Soc Jpn 44:25–42

Milliff RF, Madden RA (1996) The existence and vertical structure of fast, eastward-moving disturbances in the equatorial troposphere. J Atmos Sci 53:586–597

Moncrieff MW, Waliser DE, Miller MJ, Shapiro MA, Asrar GR, Caughey J (2012) Multiscale convective organization and the YOTC virtual global field campaign. Bull Am Meteorol Soc 93:1171–1187

Neelin JD (1988) A simple model for surface stress and low-level flow in the tropical atmosphere driven by prescribed heating. Q J Roy Meteorol Soc 114:747–770

Ogrosky HR, Stechmann SN (2015a) The MJO skeleton model with an observation-based background state and forcing. Q J Roy Meteorol Soc. https://doi.org/10.1002/qj.2552

Ogrosky HR, Stechmann SN (2015b) Assessing the equatorial long-wave approximation: asymptotics and observational data analysis. J Atmos Sci 72:4821–4843

Ogrosky HR, Stechmann SN (2016) Identifying convectively coupled equatorial waves using theoretical wave eigenvectors. Mon Weather Rev. https://doi.org/10.1175/MWR-D-15-0292.1

Stechmann SN, Majda AJ (2015) Identifying the skeleton of the Madden-Julian oscillation in observational data. Mon Weather Rev 143:395–416

Stechmann SN, Ogrosky HR (2014) Satellite observations of undamped tropical circulations. Geophys Res Lett 41:9097–9105

Straub KH (2013) MJO initiation in the real-time multivariate MJO index. J Clim 26:1130–1151

Straub KH, Kiladis GN, Ciesielski PE (2006) The role of equatorial waves in the onset of the South China Sea summer monsoon and the demise of El Niño during 1998. Dyn Atmos Oceans 42:216–238

Thual S, Majda AJ, Stechmann SN (2014) A stochastic skeleton model for the MJO. J Atmos Sci 71:697–715

Waliser DE, Moncrieff MW, Burridge D, Fink AH, Gochis D, Goswami BN, Guan B, Harr P, Heming J, Hsu H-H, Jakob C, Janiga M, Johnson R, Jones S, Knippertz P, Marengo J, Nguyen H, Pope M, Serra Y, Thorncroft C, Wheeler M, Wood R, Yuter S (2012) The "Year" of tropical convection (May 2008-April 2010). Bull Am Meteorol Soc 93:1189–1218

Webster PJ (1972) Response of the tropical atmosphere to local, steady forcing. Mon Weather Rev 100:518–541

Wheeler M, Hendon H (2004) An all-season real-time multivariate MJO index: development of an index for monitoring and prediction. Mon Weather Rev 132:1917–1932

Wheeler M, Kiladis GN (1999) Convectively coupled equatorial waves: analysis of clouds and temperature in the wavenumber-frequency domain. J Atmos Sci 56:374–399

Wheeler M, Weickmann KM (2001) Real-time monitoring and prediction of modes of coherent synoptic to intraseasonal tropical variability. Mon Weather Rev 129:2677–2694

Wheeler M, Kiladis GN, Webster PJ (2000) Large-scale dynamical fields associated with convectively coupled equatorial waves. J Atmos Sci 57:613–640

Chapter 6
Refined Vertical Structure in the Stochastic Skeleton Model for the MJO

To a first approximation, the Madden–Julian oscillation (MJO) has a symmetry (actually, an anti-symmetry) in its vertical structure: the winds in the upper troposphere are nearly equal in magnitude and opposite in sign to the winds in the lower troposphere. In earlier chapters of the book, such a vertical structure was used as a simplifying assumption in formulating the most basic versions of the MJO skeleton model. In the present chapter, more general versions of the skeleton model are described, and they incorporate deviations from this anti-symmetric vertical structure. The underlying processes that govern the vertical structure are a set of distinct cloud types. The anti-symmetric vertical structure is associated with the most basic tropical cloud type—deep convection—which extends through the depth of the troposphere, from the boundary layer to the tropopause. In addition, other important cloud types are also present: congestus clouds populate the lower troposphere, where they help to moisten and precondition the troposphere ahead of the most active phase of deep convection, and stratiform anvil clouds are present in the upper troposphere in association with deep convective cells and mesoscale convective systems. Taken together, there is a progression in development of the cloud populations of the MJO, from congestus to deep to stratiform, which imparts a vertical tilt on the MJO structure. This chapter builds these additional cloud types into the skeleton model and is thereby able to reproduce the refinements (tilts) in the MJO vertical structure seen in nature.

6.1 Introduction

In the previous chapters, many important features of the Madden–Julian oscillation in the tropics (MJO, Madden and Julian 1971, 1994) have been discussed. In addition to those features, the MJO in nature propagates eastward with an interesting vertical structure. An illustration of this feature is shown in Fig. 6.1. The MJO

© The Author(s), under exclusive licence to Springer Nature Switzerland AG 2019
A. J. Majda et al., *Tropical Intraseasonal Variability and the Stochastic Skeleton Method*, SpringerBriefs in Mathematics of Planet Earth,
https://doi.org/10.1007/978-3-030-22247-5_6

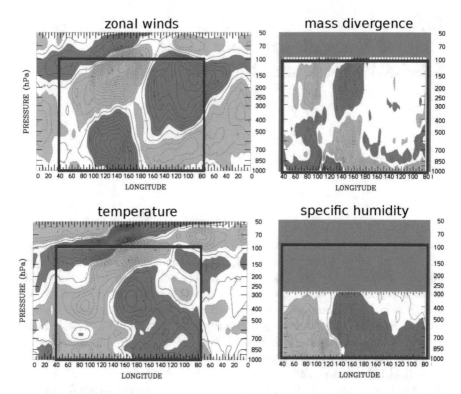

Fig. 6.1 Vertical structure of the MJO in observations, as a function of longitude (deg °E) and height (hPa). For (**a**) zonal winds (contour interval $0.5\,\mathrm{m\,s^{-1}}$), (**b**) mass divergence (contour interval $2.10^{-7}\,\mathrm{kg\,m^{-3}\,s^{-1}}$), (**c**) temperature (contour interval 0.1 K), and (**d**) specific humidity (contour interval $0.1\,\mathrm{g\,kg^{-1}}$). First, an MJO index is obtained from OLR at 155°E filtered in the intraseasonal band ($k = 0$–9 and $\omega = 1/96$–1/30 days), and, second, all fields are regressed on the index and scaled for an OLR anomaly of $-40\,\mathrm{W\,m^{-2}}$. Adapted from Kiladis et al. (2005)

envelope in nature consists of a complex front-to-rear (i.e., tilted) vertical structure as seen on all main dynamical fields such as zonal winds, divergence, temperature, and humidity (Kikuchi and Takayabu 2004; Kiladis et al. 2005; Tian et al. 2006). The front-to-rear structure of individual MJO events is often unique, with complex dynamic and convective features within the MJO envelope (e.g., westerly wind bursts, etc.) that vary from one event to another. Meanwhile, many general circulation models (GCMs) have typically had poor representations of the MJO and its main characteristics (Lin et al. 2006; Kim et al. 2009), although there has been significant recent progress in a few GCMs. The observed vertical tilt of the MJO in particular is well simulated in good GCM models but not in the poor ones, typically due to missing MJO preconditioning processes to the east of convection in the lower troposphere (Jiang et al. 2015; Jiang 2017).

Observations also reveal a central role of three cloud types above the boundary layer in the MJO: lower-middle troposphere congestus cloud decks that moisten and precondition the lower troposphere in the initial phase, followed by deep convection and a trailing wake of upper troposphere stratiform clouds. Interestingly, those three cloud types play important roles for several synoptic and mesoscale convective events for which a front-to-rear vertical structure is also observed (Moncrieff 2004; Mapes et al. 2006; Frenkel et al. 2015). Khouider and Majda (2006, 2007) developed a systematic multicloud model convective parameterization highlighting the nonlinear dynamical role of the three cloud types and their different heating vertical structures. Such a model reproduces key features of the observational record for mesoscale and synoptic-scale convectively coupled waves, and has been used to improve cumulus parameterization in GCMs (Khouider et al. 2011; Ajayamohan et al. 2013; Deng et al. 2014). Although the multicloud model focuses on the synoptic and mesoscales processes these results have important implications for understanding the characteristics of the planetary MJO structure as well. As another example, the role of synoptic scale waves in producing key features of the MJO's vertical structure has been elucidated in multiscale asymptotic models (Majda and Biello 2004; Biello and Majda 2005, 2006; Majda and Stechmann 2009a; Stechmann et al. 2013): in those models the MJO vertical tilted structure is instead a by-product (or "muscle") feature that is generated by upscale transport from the synoptic scales within the MJO envelope.

In this chapter we propose and analyze extended skeleton models accounting for the refined vertical structure of the MJO in nature. Those skeleton models are the first simple models for the MJO with dynamical tilts. They capture in particular the front-to-rear structure of moisture, winds, and temperature from Fig. 6.1 in simple fashion. This is achieved here by accounting for the three types of convective activities (congestus, deep, and stratiform) discussed above in the dynamical core of the skeleton model. During MJO events, the convective activity transits from the congestus to the deep to the stratiform type which facilitates the vertical tilted structure on other variables.

The main strategy for developing those extended skeleton models with MJO vertical tilts is to add variables in the skeleton model with suitable dynamics (Thual and Majda 2015a,b). In this chapter, we will present two extended skeleton models, either with (1) the three convective activities mentioned above or with (2) the three convective activities and two active baroclinic modes. Extended models such as (2) grasp more details of the MJO dynamics and are therefore more desirable even though their increased complexity introduces some additional caveats. In addition to this, general guidelines for designing and testing those extended skeleton models are discussed. All models follow a design prototype similar to the original skeleton model from Chap. 1 with conserved positive energy and linear neutral solutions due to the absence of dissipative processes as guiding mathematics principles. In addition, all models are tested with a stochastic parameterization of convective activity as in the previous chapters, where we analyze main characteristics of the simulated MJO.

The present chapter is organized as follows: In Sect. 6.2 we present two examples of extended skeleton models that reproduce the MJO dynamical tilts. In Sect. 6.3 we discuss general guidelines for designing and testing such types of extended skeleton models. Section 6.4 is a short discussion with concluding remarks.

6.2   Skeleton Models with Refined Vertical Structure

6.2.1   Representation of Convective Activities

This section illustrates the main features of skeleton models with refined vertical structure (Thual and Majda 2015a,b). In order to derive those extended models, let us first refine the vertical structure of the original skeleton model with several envelopes of convective activity instead of a single one, namely congestus activity a_c, deep activity a_d, and stratiform activity a_s, as shown in Fig. 6.2. Each type of convective activity has different location in the troposphere ($0 \leq z \leq 15$ km). Typically, congestus activity a_c accounts for lower-middle troposphere congestus cloud decks that moisten and heat the lower troposphere, for example, in the initial phase of the MJO. Deep activity a_d represents deep convection that heats and dries the entire troposphere. Finally, stratiform activity a_s accounts for upper troposphere stratiform clouds, for example, in the trailing wake of the MJO. Those three types of convective activities follow the same prototype as in the multicloud model (Khouider and Majda 2006, 2007; Frenkel et al. 2015), though an important distinction is that they are planetary envelopes defined on the planetary scale (instead of the usual synoptic or mesoscale that characterizes convective events). With the decomposition of convective activity into three types, the skeleton model from the previous chapters is modified as:

$$
\begin{aligned}
&\partial_t u_1 - y v_1 - \partial_x \theta_1 = 0\\
&y u_1 - \partial_y \theta_1 = 0\\
&\partial_t \theta_1 - (\partial_x u_1 + \partial_y v_1) = \overline{H}(\xi_c a_c + \xi_d a_d + \xi_s a_s) - s_1^\theta\\
&\partial_t q + \overline{Q}(\partial_x u_1 + \partial_y v_1) = -\overline{H}(\xi_c a_c + \xi_d a_d + \xi_s a_s) + s^q,
\end{aligned}
\tag{6.1}
$$

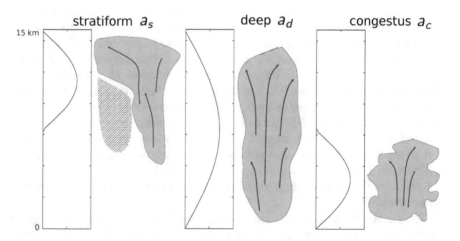

Fig. 6.2 Skeleton model with three convective activities. Schematic representation of three types of convective activity and their vertical structures, from the surface ($z = 0$) to the tropopause ($z = 15$ km)

where ξ_{1c}, ξ_{1d}, and ξ_{1s} are projection coefficients (see hereafter). Note that each convective activity heats and dries the atmosphere at the same time.

6.2.2   Evolution of Convective Activities

As shown above in Eq. (6.1), we have introduced new types of convective activities in the skeleton model. Those convective activities follow the same prototype as in the previous work (e.g., Khouider and Majda 2006, 2007); however, an important distinction is that they are planetary envelopes for which suitable evolution and interactions must be proposed. The evolution of each type of convective activity is proposed to be as follows:

$$\begin{aligned}
\partial_t a_c &= \Gamma_c(\xi_c q + \beta_s r_s - \beta_c r_d)(a_c - \epsilon_c r_c) \\
\partial_t a_d &= \Gamma_d(\xi_d q + \beta_c r_c - \beta_d r_s)(a_d - \epsilon_d r_d) \\
\partial_t a_s &= \Gamma_s(\xi_s q + \beta_d r_d - \beta_s r_c)(a_s - \epsilon_s r_s),
\end{aligned} \tag{6.2}$$

where $r_d = a_d - \bar{a}_d$, $r_c = a_c - \bar{a}_c$, and $r_s = a_s - \bar{a}_s$ are convective activity anomalies to the RCE state.

Figure 6.3 summarizes the relationships from Eq. (6.2). Those relationships are chosen based on several criteria. First, they are similar in essence to the ones of the original skeleton model, with a growth/decay of convective activities that depends on environmental conditions. For instance, in Eq. (6.2), each convective activity grows/decays (with respective rates Γ_d, Γ_c, Γ_s) depending on environmental conditions, namely a combination of moisture anomalies (q) and anomalies of the other convective activities (r_c, r_d, r_s). The environment moisture q favors

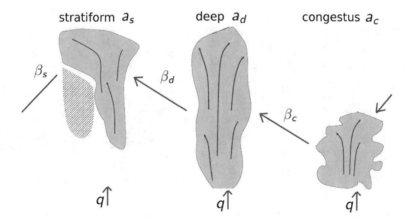

Fig. 6.3 Skeleton model with three convective activities. Interactions between the three types of convective activity

the growth/decay of deep convective activity a_d (Kikuchi and Takayabu 2004; Majda and Stechmann 2009b), and by extension we assume that it also favors the growth/decay of congestus and stratiform activity, a_c and a_s. Second, the relationships in Eq. (6.2) are consistent with empirical relationships between different convective activities found elsewhere (e.g., in the multicloud model, see Khouider and Majda 2006; Khouider et al. 2010). An environment with enhanced/suppressed convective activity of a certain type favors the growth/decay of another type of convective activity, with transition rates β_c, β_d, β_s. For example, enhanced congestus activity a_c ($r_c \geq 0$) favors the growth of deep activity a_d in the second row of Eq. (6.2), while inversely enhanced deep activity ($r_d \geq 0$) favors the decay of congestus activity in the first row of Eq. (6.2), both with transition rate β_c. Finally, we introduce some pseudo-dissipation coefficients $0 \leq \epsilon_c, \epsilon_d, \epsilon_s \leq 1$: for more realism, we assume that the growth of deep activity a_d is near exponential ($\epsilon_d = 0$), while the growth of congestus and stratiform activity is near linear ($\epsilon_c, \epsilon_s = 0.9$).

6.2.3   Skeleton Model with Three Convective Activities

The skeleton model with three convective activities (Eqs. (6.1)–(6.2)) has an MJO linear solution with a tilted vertical structure, as shown in Fig. 6.4 (Thual and Majda 2015b). As in the original skeleton model, we retrieve moisture anomalies q leading convective activity a. In addition to this, the present model reproduces the front-to-rear structure of the MJO observed in nature on all fields such as heating, moisture, temperature, and winds (Kiladis et al. 2005). The front-to-rear structure is oriented westward/upward, with a progressive transition from congestus to deep to stratiform activity, a transition from lower to middle level moisture anomalies, and a transition

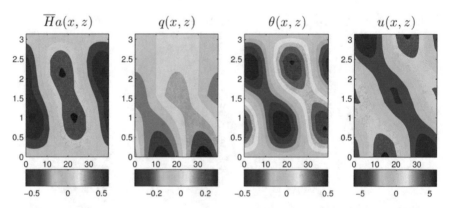

Fig. 6.4 Skeleton model with three convective activities. Structure $x-z$ of the MJO linear solution at the equator for $k = 1$. Contours of reconstructed heating $\overline{H}a$ (K day^{-1}), moisture q (K), potential temperature θ (K), and zonal winds u (ms^{-1}), at the equator and as a function of x (1000 km) and z ($0 \leq z \leq \pi$ from the bottom to the tropopause)

from lower to upper wind anomalies and lower to upper temperature anomalies. This is an attractive feature of the present skeleton model. In order to retrieve the tilted structure in Fig. 6.4 on all variables a secondary slaved circulation must be considered, as described later in Sect. 3.4.

The skeleton model with three convective activities also captures realistic features of the MJO discussed in the previous chapters, such as power spectra and intermittent MJO wave trains (Thual and Majda 2015b). For this we replace the deterministic evolutions of convective activities in Eq. (6.2) with corresponding stochastic birth/death processes (as in Chap. 3). This accounts for the irregular and intermittent contribution of unresolved synoptic convective activity to the intraseasonal–planetary dynamics. Figure 6.5 shows the details of a selected MJO wave train for solutions of such a stochastic skeleton model. In particular, congestus activity $\overline{H}a_c$ and moisture q lead the deep convective activity center $\overline{H}a_d$, while stratiform activity $\overline{H}a_s$ trails behind, consistent with the MJO front-to-rear structure of convective activity from Fig. 6.4. Figure 6.6 shows a power spectra of the variables as a function of the zonal wavenumber k (in $2\pi/40,000$ km) and frequency ω (in cpd). The present skeleton model with refined vertical structure simulates an MJO-like signal that is the dominant signal at intraseasonal–planetary scale, consistent with observations (Wheeler and Kiladis 1999) and the previous chapters. The MJO appears here as a sharp power peak in the intraseasonal–planetary band ($1 \leq k \leq 5$ and $1/90 \leq \omega \leq 1/30$ cpd), most prominent in u_1, q, $\overline{H}a_d$, and also $\overline{H}a_c$ and $\overline{H}a_s$. Due to the increased complexity of the present skeleton

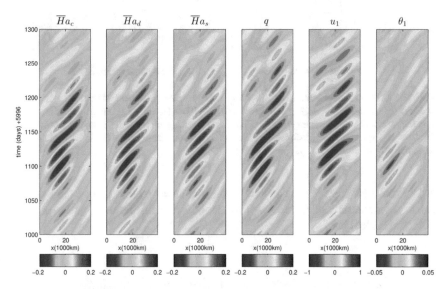

Fig. 6.5 Skeleton model with three convective activities. Hovmollers $x - t$ for a selected MJO wave train. For $\overline{H}a_c$, $\overline{H}a_d$, $\overline{H}a_s$ (K day^{-1}), q (K), u_1 (ms^{-1}), and θ_1(K) at the equator, as a function of zonal position (1000 km) and simulation time (days from an arbitrary reference time). All variables are filtered in the MJO band ($k = 1$–3, $\omega = 1/30$–$1/70$ cpd)

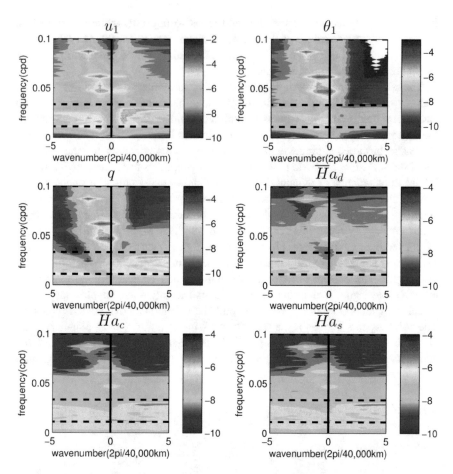

Fig. 6.6 Skeleton model with three convective activities. Zonal wavenumber–frequency power spectra: for u_1 (ms^{-1}), θ_1 (K), q (K), $\overline{H}a_d$, $\overline{H}a_c$, and $\overline{H}a_s$(K day^{-1}) taken at the equator, as a function of zonal wavenumber (in $2\pi/40{,}000$ km) and frequency (cpd). The contour levels are in the base 10-logarithm, for the dimensional variables taken at the equator. The black dashed lines mark the periods 90 and 30 days

model, the MJO power peak is here less prominent than the one of the original stochastic skeleton model (see Chap. 3), but still remains the main power peak in the intraseasonal band.

6.2.4   Skeleton Model with Two Active Baroclinic Modes

The front-to-rear vertical structure of the MJO in nature is observed on many variables such as heating, moisture, winds, temperature, etc. (e.g., Kiladis et al.

2005). In the above skeleton model, such a vertical structure is accurately captured, thanks to the introduction of three convective activities (Eqs. (6.1)–(6.2)). However, an important part of the circulation in that model such as the second baroclinic mode dynamics is slaved instead of dynamically coupled.

More dynamical details of the MJO vertical structure can be added to the skeleton model, following a systematic formalism described in the next sections. As an example we add here coupling to the second baroclinic mode dynamics (Thual and Majda 2015a). The moisture budget from Eq. (6.1) in particular is modified as:

$$\partial_t q + \overline{Q}(\partial_x u_1 + \partial_y v_1) + \lambda \overline{Q}(\partial_x u_2 + \partial_y v_2)/2$$

$$= -\overline{H}(\xi_{1c}a_c + \xi_{1d}a_d + \xi_{1s}a_s) + \lambda \overline{H}(\xi_{2s}a_s - \xi_{2c}a_c) + s^q \qquad (6.3)$$

which includes additional moisture convergence $\lambda \overline{Q}(\partial_x u_2 + \partial_y v_2)$ due to the second baroclinic mode (with zonal winds u_2, v_2). For consistency with the skeleton model dynamics there are in addition modifications to the moistening rates (ξ_{2c}, ξ_{2s}) as well as the amplitude equations (the details of those modifications are shown in the next sections). This extended model is more realistic than the former one and grasps more details of the MJO tilted structure: for instance, there are now fully coupled interactions between the planetary-scale dry dynamics of the first and second baroclinic modes.

The features of such a skeleton model with two active baroclinic modes are consistently similar to the ones shown previously. First, a tilted MJO linear solution is found as in Fig. 6.4 (not shown). Second, the power spectra and intermittent MJO wave trains are also captured. Figure 6.7 shows an example of an MJO wave train, similar to the one in Fig. 6.5 but now including additional dynamics of the second baroclinic mode (u_2, θ_2). Figure 6.8 shows the associated power spectra, similar to the ones in Fig. 6.6 with an MJO signal that is dominant at intraseasonal–planetary scale. Although the MJO vertical structure is more realistic in this model the MJO signal is slightly less prominent, notably because of the additional variables that introduce new signals (around $\omega \approx 0.01$ *cpd*).

6.3 Model Formulation

This section details the mathematical formulation of extended skeleton models with refined vertical structure. Two examples of those models have been presented in the previous section, either (1) with three convective activities or (2) with three convective activities and two active baroclinic modes. Here general guidelines for designing and testing those extended skeleton models are discussed. All models follow a prototype similar to the original skeleton model in Chap. 2 with conserved positive energy and linear neutral solutions due to the absence of dissipative processes.

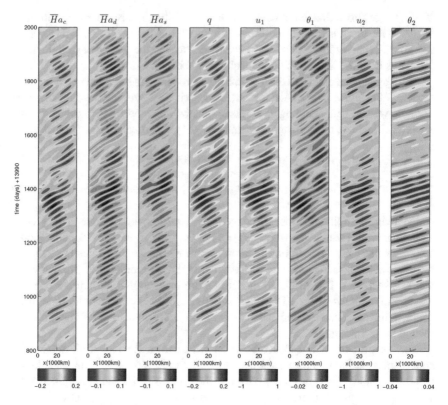

Fig. 6.7 Skeleton model with two active baroclinic modes: Hovmollers ($x - t$ diagrams) at equator filtered in the MJO band (here k = 1–3, w = 1/30–1/70 cpd)

6.3.1   Vertical Baroclinic Structure

We first briefly recall the characteristics of the vertical structure in the original skeleton model (see Chap. 2). First recall that the starting skeleton model reads, in nondimensional units,

$$
\begin{aligned}
\partial_t u - yv &= -\partial_x p \\
yu &= -\partial_y p \\
0 &= -\partial_z p + \theta \\
\partial_x u + \partial_y v + \partial_z w &= 0 \\
\partial_t \theta + w &= \overline{H}a - s^\theta \\
\partial_t q - \overline{Q}w &= -\overline{H}a + s^q \\
\partial_t a &= \Gamma qa \,,
\end{aligned}
\tag{6.4}
$$

with periodic boundary conditions along the equatorial belt. The five first rows of Eq. (6.4) describe the dry atmosphere dynamics, with equatorial long-wave scaling

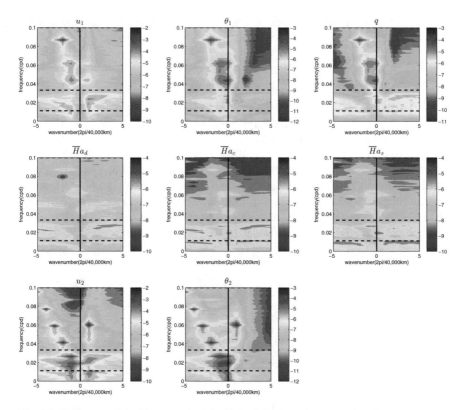

Fig. 6.8 Skeleton model with two active baroclinic modes: zonal wavenumber–frequency power spectra. See Fig. 6.6 for definitions

as allowed at planetary scale. The u, v, and w are the zonal, meridional, and vertical velocity, respectively; and p and θ are the pressure and potential temperature, respectively. The sixth row describes the evolution of low-level moisture q, and the seventh row is the nonlinear amplitude equation for a described in the previous chapters. All variables are anomalies from a radiative–convective equilibrium, except a with $a \geq 0$.

Next, the system from Eq. (6.4) is projected and truncated to the first vertical structures. For this flow within the equatorial troposphere ($0 \leq z \leq \pi$ in nondimensional units) the relevant structures are the first few vertical baroclinic modes (Majda 2003), with a reconstruction of total fields that reads:

$$
\begin{aligned}
\{u, v, p\}(x, y, z, t) &= \sum_n \{u_n, v_n, p_n\} F_n(z) \\
\{\theta, w, s^\theta\}(x, y, z, t) &= \sum_n \{\theta_n, w_n, s_n^\theta\} G_n(z),
\end{aligned}
\tag{6.5}
$$

where $n \geq 1$ for the first few baroclinic modes with $F_n(z) = \sqrt{2}\cos(nz)$ and $G_n(z) = \sqrt{2}\sin(nz)$, for $0 \leq z \leq \pi$. The original skeleton model from Chaps. 1,

2, 3 has a simple and coarse vertical structure. This is obtained by truncation to the first baroclinic mode $n = 1$, which reads:

$$
\begin{aligned}
&\partial_t u_1 - y v_1 - \partial_x \theta_1 = 0 \\
&y u_1 - \partial_y \theta_1 = 0 \\
&\partial_t \theta_1 - (\partial_x u_1 + \partial_y v_1) = \overline{H} a - s_1^\theta \\
&\partial_t q + \overline{Q}(\partial_x u_1 + \partial_y v_1) = -\overline{H} a + s^q \\
&\partial_t a = \Gamma q a \,,
\end{aligned}
\tag{6.6}
$$

where the dry dynamics component is now a time-dependent and non-dissipative version of the Matsuno–Gill model (Matsuno 1966; Gill 1980). In order to keep the system simple the vertical structure of lower-level moisture q and convective a is not made explicit, even though the interaction of those variables with the first baroclinic mode $n = 1$ is straightforward to understand.

6.3.2 Coupling to Three Convective Activities

The starting skeleton model from Eq. (6.6) is now extended to include three types of convective activities, a_d, a_c, and a_s. This reads, in nondimensional units:

$$
\begin{aligned}
&\partial_t u_1 - y v_1 - \partial_x \theta_1 = 0 \\
&y u_1 - \partial_y \theta_1 = 0 \\
&\partial_t \theta_1 - (\partial_x u_1 + \partial_y v_1) = \overline{H}(\xi_{1c} a_c + \xi_{1d} a_d + \xi_{1s} a_s) - s_1^\theta \\
&\partial_t q + \overline{Q}(\partial_x u_1 + \partial_y v_1) = -\overline{H}(\xi_{1c} a_c + \xi_{1d} a_d + \xi_{1s} a_s) + s^q \\
&\partial_t a_c = \Gamma_c (\xi_{1c} q + \beta_s r_s - \beta_c r_d)(a_c - \epsilon_c r_c) \\
&\partial_t a_d = \Gamma_d (\xi_{1d} q + \beta_c r_c - \beta_d r_s)(a_d - \epsilon_d r_d) \\
&\partial_t a_s = \Gamma_s (\xi_{1s} q + \beta_d r_d - \beta_s r_c)(a_s - \epsilon_s r_s) \,.
\end{aligned}
\tag{6.7}
$$

This model is identical to the one discussed in the previous section in Eq. (6.1)–(6.2) and in Figs. 6.4, 6.5, 6.6. In addition, for $a_d = a$, $\xi_{1d} = 1$, $\epsilon_d = 0$, and $a_c, a_s = 0$ we retrieve the original skeleton model from Eq. (6.6).

For simplicity, we associate simple vertical structures to each type of convective activity that are consistent with the skeleton model. Those simplified vertical structures are shown in Fig. 6.9: the a_d, a_c, and a_s are associated with the structures $F_d(z)$, $F_c(z)$, and $F_s(z)$, respectively. We consider half-sinusoids $F_d = G_1$, $F_c = G_2$ for $0 \le z \le \pi/2$ and $F_s = -G_2$ for $\pi/2 \le z \le \pi$. The total convective activity decomposes as $a(x, y, z, t) = a_c F_c(z) + a_d F_d(z) + a_s F_s(z)$, and is always positive. In Eq. (6.7) we consider the dynamics of the first baroclinic mode, and therefore introduce projection coefficients of convective activity on the first baroclinic mode, ξ_{1d}, ξ_{1c}, and ξ_{1s}. The projection coefficients ξ_{1d}, ξ_{1s}, and ξ_{1s} have values $\xi_{1d} = 1$ and $\xi_{1c}, \xi_{1s} = 4/3\pi$. Note that $\xi_{1d} \ge \xi_{1c}, \xi_{1s}$, therefore a unit of deep activity heats/dries the first baroclinic mode more than

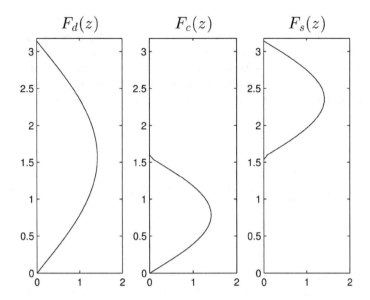

Fig. 6.9 Vertical structures $F_d(z)$, $F_c(z)$, and $F_s(z)$ in the skeleton model with three convective activities, as a function of z ($0 \leq z \leq \pi$ from the bottom to the tropopause)

a unit of congestus or stratiform activity. Different values of those projection coefficients could however be considered assuming a more complex localization of heating/drying by convective activity, for example, a decreased ξ_{1s} due to radiative effects, etc. (see, e.g., Khouider and Majda 2008; Frenkel et al. 2015 for a discussion).

6.3.3   Conserved Energy Principle

The above skeleton model with three convective activities in Eq. (6.7) conserves a total positive energy (as there are no dissipative processes). This reads:

$$\partial_t[E_1 + E_Z + E_c + E_d + E_s] + A_1 = 0$$
$$E_1 = (\overline{Q}/\overline{H})[\tfrac{1}{2}u_1^2 + \tfrac{1}{2}\theta_1^2]$$
$$E_Z = (\overline{Q}/\overline{H})[\tfrac{1}{2}(\overline{Q}(1-\overline{Q}))^{-1}(q_1 + \overline{Q}\theta_1)^2]$$
$$A_1 = (\overline{Q}/\overline{H})[-\partial_x(u_1\theta_1) - \partial_y(v_1\theta_1)] \quad\quad (6.8)$$
$$E_c = \Gamma_c^{-1}(1-\epsilon_c)^{-2}(a_c - \epsilon_c a_c - \overline{a}_c \log(a_c))$$
$$E_d = \Gamma_d^{-1}(1-\epsilon_d)^{-2}(a_d - \epsilon_d a_d - \overline{a}_d \log(a_d))$$
$$E_s = \Gamma_s^{-1}(1-\epsilon_s)^{-2}(a_s - \epsilon_s a_s - \overline{a}_s \log(a_s)).$$

The above condition of a conserved energy is an important requisite for the skeleton model to exhibit neutral solutions. This condition imposes several aspects of Eq. (6.7): for example, each convective activity heats and dries the atmosphere at the same time and grows with positive moisture anomalies, and the transition rate terms $(\beta_c, \beta_d, \beta_s)$ are skew symmetric terms (that cancel in the energy budget). Note also that the model conserves a vertically integrated moist static energy:

$$\partial_t(\theta_1 + q) - (1 - \overline{Q})(\partial_x u_1 + \partial_y v_1) = 0. \tag{6.9}$$

6.3.4   Addition of a Slaved Secondary Circulation

In addition to the above dynamical core of the skeleton model with three convective activities, we consider a slaved secondary circulation in order to grasp more details of the vertical structure. This allows us to retrieve the tilted vertical structure on all variables in Fig. 6.4 including moisture, temperature, and zonal winds. For instance, a breakdown of the MJO linear solution from Fig. 6.4 is shown in Fig. 6.10 that

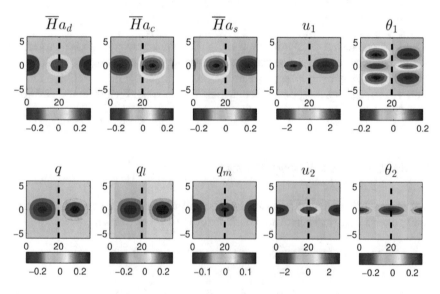

Fig. 6.10 Model with three convective activities. Structure $x - y$ of the MJO mode for $k = 1$, in dimensional units: $\overline{H}a_d$, $\overline{H}a_c$, and $\overline{H}a_s$ (K day^{-1}), u_1(ms^{-1}), θ_1(K), q, q_l, and q_m (K), u_2 (ms^{-1}), θ_2 (K), as a function of x and y (1000 km). The dashed line marks $x = 20,000$ km

includes the additional fields from the secondary circulation. First, we consider the
dry dynamics of the second baroclinic mode:

$$\partial_t u_2 - y v_2 - \partial_x \theta_2/2 = 0$$
$$y u_2 - \partial_y \theta_2/2 = 0 \tag{6.10}$$
$$\partial_t \theta_2 - (\partial_x u_2 + \partial_y v_2)/2 = \overline{H}(\xi_{2c} a_c - \xi_{2s} a_s) - s_2^\theta \,,$$

where u_n, v_n, and θ_n are the zonal, meridional velocity, and potential temperature
anomalies, respectively, for the baroclinic mode $n = 2$. The s_2^θ is an external source
of cooling, and ξ_{2c}, ξ_{2s} are projection coefficients. Second, we consider a more
refined vertical structure of moisture. We assume here that q is a vertically integrated
estimate of moisture that decomposes into $q = q_l + \alpha q_m$, where q_l is moisture
anomalies at bottom level and q_m at middle level (here at $z = \pi/4$ and $z = \pi/2$,
respectively, though other values could be considered) and $\alpha \geq 0$ measures the
middle level moisture contribution to q. This slightly extends the moisture definition
of the original skeleton model (for which $\alpha = 0$) because moisture leading the MJO
convective core may sometimes have contributions from different levels (Stechmann
and Majda 2015). We assume that the evolution of lower and middle tropospheric
moisture anomalies is as follows:

$$\partial_t q_m = M(a_c - a_s) + s_m^q$$
$$\partial_t q_l + \overline{Q}(\partial_x u_1 + \partial_y v_1) = -\overline{H}(\xi_{1c} a_c + \xi_{1d} a_d + \xi_{1s} a_s) - \alpha M(a_c - a_s) + s_l^q \,, \tag{6.11}$$

where s_m^q and s_l^q are external sources of moistening satisfying $s^q = s_l^q + \alpha s_m^q$.
In addition to the main drying and moisture convergence occurring at lower level,
congestus and stratiform activity favors moisture exchange between the lower and
middle level at a rate M (e.g., through detrainment or downdrafts, see Khouider and
Majda 2006). Noteworthy, the above equations sum up to the moisture budget of
$q = q_l + \alpha q_m$ in Eq. (6.7).

6.3.5 Active Coupling to the Second Baroclinic Mode

The main motivation for the new classes of skeleton models is to add processes to
the original skeleton model in order to capture more dynamical details of the MJO
tilted structure in nature. Let us now modify the skeleton model such that the second
baroclinic mode is dynamically coupled (instead of slaved as in Eq. (6.10)). With
this modification we obtain exactly the skeleton model from the previous section as
in Eq. (6.3) and Figs. 6.7 and 6.8 with two active baroclinic modes. This reads, in
nondimensional units:

$$\partial_t u_n - y v_n - \partial_x \theta_n / n = 0$$
$$y u_n - \partial_y \theta_n / n = 0$$
$$\partial_t \theta_1 - (\partial_x u_1 + \partial_y v_1) = \overline{H}(\xi_{1d} a_d + \xi_{1c} a_c + \xi_{1s} a_s) - s_1^\theta$$
$$\partial_t \theta_2 - (\partial_x u_2 + \partial_y v_2)/2 = \overline{H}(\xi_{2c} a_c - \xi_{2s} a_s) - s_2^\theta$$
$$\partial_t q + \overline{Q}(\partial_x u_1 + \partial_y v_1) + \lambda \overline{Q}(\partial_x u_2 + \partial_y v_2)/2 = -\overline{H}(\xi_{1c} a_c + \xi_{1d} a_d + \xi_{1s} a_s)$$
$$\qquad + \lambda \overline{H}(\xi_{2s} a_s - \xi_{2c} a_c) + s^q$$
$$\partial_t a_c = \Gamma_c[\,(\xi_{1c} + \lambda \xi_{2c}) q + (\lambda \xi_{2c} \overline{Q}) \theta_1 + (\lambda \xi_{1c} \overline{Q}) \theta_2 + \beta_s r_s - \beta_c r_d\,](a_c - \epsilon_c r_c)$$
$$\partial_t a_d = \Gamma_d[\,\xi_{1d} q + (\lambda \xi_{1d} \overline{Q}) \theta_2 + \beta_c r_c - \beta_d r_s\,](a_d - \epsilon_d r_d)$$
$$\partial_t a_s = \Gamma_s[\,(\xi_{1s} - \lambda \xi_{2s}) q - (\lambda \xi_{2s} \overline{Q}) \theta_1 + (\lambda \xi_{1s} \overline{Q}) \theta_2 + \beta_d r_d - \beta_s r_c\,](a_s - \epsilon_s r_s).$$
$$\tag{6.12}$$

The above modifications follow an important prototype based on the conservation of a positive energy. Initially, we only wanted to add a single additional term in the model, namely moisture convergence of the second baroclinic mode $\lambda \overline{Q}(\partial_x u_2 + \partial_y v_2)/2$ with arbitrary intensity λ (in the fifth row of Eq. (6.12)). However, as shown above in Eq. (6.12) there are instead many additional terms such as additional moistening rates (with parameters ξ_{2s}, ξ_{2c}) as well as additional growth/decay rates (proportional to $\lambda \theta_1$ or $\lambda \theta_2$) for convective activities. The reason for this is that we ensure that the skeleton's model positive energy is still conserved when adding additional processes. The energy budget associated with Eq. (6.12) for balanced external sources ($s_1^\theta + \lambda s_2^\theta = s^q$) is similar to the one from Eq. (6.8) with a few additional or modified terms:

$$\partial_t[E_1 + E_2 + E_Z + E_c + E_d + E_s] + A_1 + A_2 = 0$$
$$E_Z = (\overline{H}(1 - \overline{Q}))^{-1}[\tfrac{1}{2}(q + \overline{Q} \theta_1 + \lambda \overline{Q} \theta_2)^2]$$
$$E_2 = (\lambda^2 \overline{Q}/\overline{H})[\tfrac{1}{2} u_2^2 + \tfrac{1}{2} \theta_2^2]$$
$$A_2 = (\lambda^2 \overline{Q}/\overline{H})[-\partial_x (u_2 \theta_2)/2 - \partial_y (v_2 \theta_2)/2].$$
$$\tag{6.13}$$

In particular, all additional terms are proportional to λ^2 and balance each other. This energy conservation principle shapes the form of the skeleton model from Eq. (6.12). Note that the model conserves a vertically integrated moist static energy, now with contribution from both the first and second baroclinic mode:

$$\partial_t (q + \theta_1 + \lambda \theta_2) - (1 - \overline{Q})(\partial_x u_1 + \partial_y v_1) - \lambda (1 - \overline{Q})(\partial_x u_2 + \partial_y v_2)/2 = 0. \tag{6.14}$$

6.3.6   A Suite of Extended Skeleton Models

The general guidelines discussed above for designing and testing extended skeleton models can be easily used. Table 6.1 shows an intercomparison of main model features between the original skeleton model (Majda and Stechmann 2009b), the two extended skeleton models discussed in this chapter (Thual and Majda 2015a,b), and an additional extended skeleton model analyzed in Thual and Majda (2015a).

Table 6.1 Highlight and intercomparison of extended skeleton models with refined vertical structures

Model	Original skeleton	Three convective activities	Two active baroclinic modes	Coupled lower and middle level moisture
Publication	Majda and Stechmann (2009b)	Thual and Majda (2015b)	Thual and Majda (2015a)	Thual and Majda (2015a)
MJO front-to-rear structure	No	Yes	Yes	Yes
Conserved positive energy	Yes	Yes	Yes	Yes
Conserved moist static energy	Yes	Yes	Yes	Yes
Coupled circulation	u_1,v_1,θ_1, q,a	$u_1,v_1,\theta_1,$ q,a_d,a_c,a_s	$u_1,v_1,\theta_1,$ $u_2,v_2,\theta_2,$ q,a_d,a_c,a_s	$u_1,v_1,\theta_1,$ $u_2,v_2,\theta_2,$ q_l,q_m,a_d,a_c,a_s
Secondary slaved circulation	Undefined	u_2,v_2,θ_2,q_l,q_m	q_l, q_m	None
Additional linear modes	None	Slow eastward Slow westward	Slow eastward Slow westward Dry Kelvin $n = 2$ Dry Rossby $n = 2$	Slow eastward Slow westward Zonally symmetric Dry Kelvin $n = 2$ Dry Rossby $n = 2$
Prominent MJO variability	Yes	Yes	Yes	No

The general guidelines are depicted in Table 6.1. First, as compared to the original skeleton model all extended models capture the MJO vertical tilted structure. Second, all extended models follow a prototype similar to the original skeleton model with conserved positive energy ensuring that the intraseasonal–planetary dynamics remain neutrally stable. Optionally those models may also conserve a vertically integrated moist static energy. Third, as models increase in complexity they add new coupled variables that grasp more dynamical details of the MJO tilted structure. Coupling those variables removes the necessity of a secondary slaved circulation, which is potentially more realistic. However, a drawback of this higher dimensionality is the introduction of several additional linear solutions (for example, a slow eastward and westward mode, a zonally symmetric mode, etc.) that must be systematically categorized and may sometimes interfere with the MJO variability itself. Finally, all models are tested with a stochastic parameterization of convective activity as in the previous chapters, where we analyze the characteristics of the MJO power spectra and wave trains: those tests conclude if a prominent and realistic MJO variability is still simulated in the extended skeleton models. In Table 6.1, for example, all models satisfy the guidelines discussed above; however, the extended skeleton model with coupled lower and middle level moisture does not show a prominent MJO variability.

6.4   Discussion

In this chapter we have proposed and analyzed several skeleton models for the MJO with refined vertical structure (Thual and Majda 2015a,b). Those skeleton models are the first simple models for the MJO with dynamical tilts. They reproduce qualitatively the front-to-rear (i.e., tilted) vertical structure of the MJO found in nature, with MJO events marked by a planetary envelope of convective activity transitioning from the congestus to the deep to the stratiform type, in addition to a front-to-rear structure of heating, moisture, winds, and temperature.

A general question that this chapter addresses is to which extent the skeleton model can be complexified toward more realism while conserving its important design features. As compared to the original skeleton model, the extended models proposed here all have their own strengths and weaknesses. They capture more details of the MJO vertical structure and have richer dynamics. They couple additional variables based on different strategies, for example, the detailed evolution of convective activity or the dynamics of the second baroclinic mode. In this chapter, we have presented two extended skeleton models showing MJO vertical tilts, either (1) with the three convective activities mentioned above or (2) with the three convective activities and two active baroclinic modes. Extended models such as (2) grasp more details of the MJO dynamics and are therefore more desirable even though their increased complexity introduces some additional caveats. A drawback of the higher dimensionality is, for example, the introduction of several additional linear solutions that may interfere with the MJO variability. Recall that a general assumption in the original skeleton model from the previous chapters (Majda and Stechmann 2009b) and the present models is that the MJO arises from neutrally stable interactions at the planetary scale, while main instabilities occur on the synoptic scale. This view differs from the one of an MJO planetary instability (Zhang 2005), and is also the reason why additional linear solutions (also neutrally stable) must be treated carefully as they may be as prominent as the MJO variability. Additional classes of skeleton models that respect this assumption may be constructed starting from the energy conservation principle introduced in this chapter.

In addition to this, the present simple models may be used to analyze zonal variations of the characteristics of the MJO front-to-rear structure (Kiladis et al. 2005) or to provide new theoretical estimates of MJO events in observations (Stechmann and Majda 2015). More complete models should also account for more detailed sub-planetary processes within the envelope of the MJO, including, for example, synoptic-scale convectively coupled waves and/or mesoscale convective systems that also affect the MJO vertical structure (e.g., Moncrieff et al. 2007; Majda et al. 2007; Khouider et al. 2010; Frenkel et al. 2012).

References

Ajayamohan S, Khouider B, Majda AJ (2013) Realistic initiation and dynamics of the Madden-Julian Oscillation in a coarse resolution aquaplanet GCM. Geophys Res Lett 40:6252–6257

Biello JA, Majda AJ (2005) A New Multiscale Model for the Madden-Julian Oscillation. J Atmos Sci 62(6):1694–1721. https://doi.org/10.1175/JAS3455.1, http://journals.ametsoc.org/doi/abs/10.1175/JAS3455.1

Biello JA, Majda AJ (2006) Modulating synoptic scale convective activity and boundary layer dissipation in the IPESD models of the Madden-Julian oscillation. Dyn Atmos Oceans 42(1–4):152–215. https://doi.org/10.1016/j.dynatmoce.2005.10.005, http://adsabs.harvard.edu/abs/2006DyAtO..42..152B

Deng Q, Khouider B, Majda AJ (2014) The MJO in a Coarse-Resolution GCM with a stochastic multicloud parameterization. J Atmos Sci 72:55–74

Frenkel Y, Majda AJ, Khouider B (2012) Using the stochastic multicloud model to improve tropical convective parameterization: a paradigm example. J Atmos Sci 69:1080–1105. https://doi.org/10.1175/JAS-D-11-0148.1

Frenkel Y, Majda A, Stechmann S (2015) Cloud-radiation feedback and atmosphere-ocean coupling in a stochastic multicloud mode. Dyn Atmos Oceans 71:35–55

Gill A (1980) Some simple solutions for heat-induced tropical circulation. Q J Roy Meteorol Soc 106:447–462

Jiang X (2017) Key processes for the eastward propagation of the Madden-Julian Oscillation based on multi-model simulations. J Geophys Res Atmos 122:755–770. https://doi.org/10.1002/2016JD025955

Jiang X et al (2015) Vertical structure and diabatic processes of the Madden-Julian Oscillation: exploring key model physics in climate simulations. J Geophys Res Atmos 120:4718–4748

Khouider B, Majda AJ (2006) A simple multicloud parameterization for convectively coupled tropical waves. Part I: linear analysis. J Atmos Sci 63:1308–1323

Khouider B, Majda AJ (2007) A simple multicloud parameterization for convectively coupled tropical waves. Part II. Nonlinear simulations. J Atmos Sci 64:381–400

Khouider B, Majda AJ (2008) Multicloud models for organized tropical convection: enhanced congestus heating. J Atmos Sci 65(3):895–914. https://doi.org/10.1175/2007JAS2408.1, http://journals.ametsoc.org/doi/abs/10.1175/2007JAS2408.1

Khouider B, Biello JA, Majda AJ (2010) A stochastic multicloud model for tropical convection. Commun Math Sci 8(1):187–216

Khouider B, St-Cyr A, Majda AJ, Tribbia J (2011) The MJO and convectively coupled waves in a coarse-resolution GCM with a simple multicloud parameterization. J Atmos Sci 68(2):240–264. https://doi.org/10.1175/2010JAS3443.1, http://journals.ametsoc.org/doi/abs/10.1175/2010JAS3443.1

Kikuchi K, Takayabu YN (2004) The development of organized convection associated with the MJO during TOGA COARE IOP: trimodal characteristics. Geophys Res Lett 31. https://doi.org/10.1029/2004GL019601

Kiladis GN, Straub KH, Haertel PT (2005) Zonal and vertical structure of the Madden-Julian oscillation. J Atmos Sci 62:2790–2809

Kim D, Sperber K, Stern W, Waliser D, Kang IS, Maloney E, Wang W, Weickmann K, Benedict J, Khairoutdinov M, Lee MI, Neale R, Suarez M, Thayer-Calder K, Zhang G (2009) Application of MJO simulation diagnostics to climate models. J Clim 22:6413–6436

Lin JL, Kiladis GN, Mapes BE, Weickmann KM, Sperber KR, Lin W, Wheeler MC, Schubert SD, Del Genio A, Donner LJ, Emori S, Gueremy JF, Hourdin F, Rasch PJ, Roeckner E, Scinocca JF (2006) Tropical Intraseasonal Variability in 14 IPCC AR4 Climate models. Part I: convective signals. J Clim 19:2665–2690. https://doi.org/10.1175/JCLI3735.1, http://journals.ametsoc.org/doi/abs/10.1175/JCLI3735.1

Madden RE, Julian PR (1971) Detection of a 40–50 day oscillation in the zonal wind in the tropical Pacific. J Atmos Sci 28:702–708

Madden RE, Julian PR (1994) Observations of the 40–50 day tropical oscillation-a review. Mon Weather Rev 122:814–837

Majda AJ (2003) Introduction to PDEs and waves for the atmosphere and ocean. Courant lecture notes in mathematics, vol 9. American Mathematical Society, providence, x+234pp

Majda AJ, Biello JA (2004) A multiscale model for tropical intraseasonal oscillations. Proc Natl Acad Sci USA 101:4736–4741. https://doi.org/10.1073/pnas.0401034101, http://libra.msra.cn/Publication/15169461/a-multiscale-model-for-tropical-intraseasonal-oscillations

Majda AJ, Stechmann SN (2009a) A simple dynamical model with features of convective momentum transport. J Atmos Sci 66:373–392. https://doi.org/10.1175/2008JAS2805.1, http://journals.ametsoc.org/doi/abs/10.1175/2008JAS2805.1

Majda AJ, Stechmann SN (2009b) The skeleton of tropical intraseasonal oscillations. Proc Natl Acad Sci 106:8417–8422. https://doi.org/10.1073/pnas.0903367106, http://www.pnas.org/content/106/21/8417

Majda AJ, Stechmann SN, Khouider B (2007) Madden-Julian oscillation analog and intraseasonal variability in a multicloud model above the equator. Proc Natl Acad Sci USA 104:9919–9924

Mapes B, Tulich S, Lin J, Zuidema P (2006) The mesoscale convection life cycle: building block or prototype for large-scale tropical waves? Dyn Atmos Oceans 42(1–4):3–29. https://doi.org/10.1016/j.dynatmoce.2006.03.003, http://adsabs.harvard.edu/abs/2006DyAtO..42....3M

Matsuno T (1966) Quasi-geostrophic motions in the equatorial area. J Meteorol Soc Jpn 44:25–43

Moncrieff M (2004) Analytic representation of the large-scale organization of tropical convection. Q J Roy Meteorol Soc 130:1521–1538

Moncrieff MW, Shapiro M, Slingo J, Molteni F (2007) Collaborative research at the intersection of weather and climate. WMO Bull 56:204–211

Stechmann SN, Majda AJ (2015) Identifying the skeleton of the Madden-Julian oscillation in observational data. Mon Weather Rev 143:395–416

Stechmann S, Majda AJ, Skjorshammer D (2013) Convectively coupled wave-environment interactions. Theor Comput Fluid Dyn 27(3–4):513–532

Thual S, Majda A (2015a) A suite of skeleton models for the MJO with refined vertical structure. Math Clim Weather Forecast 1:70–95

Thual S, Majda AJ (2015b) A skeleton model for the MJO with refined vertical structure. Clim Dyn 46. https://doi.org/10.1007/s00382-015-2731-x

Tian B, Waliser D, Fetzer E, Lambrigsten B, Yung Y, Wang B (2006) Vertical moist thermodynamic structure and spatial-temporal evolution of the MJO in AIRS observations. J Atmos Sci 63:2462–2485

Wheeler M, Kiladis GN (1999) Convectively coupled equatorial waves: analysis of clouds and temperature in the wavenumber-frequency domain. J Atmos Sci 56:374–399. https://doi.org/10.1175/1520-0469(1999)056<0374:CCEWAO>2.0.CO;2

Zhang C (2005) Madden-Julian oscillation. Rev Geophys 43, rG2003. https://doi.org/10.1029/2004RG000158

Chapter 7
Current and Future Research Perspectives

Here we briefly discuss several interesting research directions in Tropical Intraseasonal Variability (TISV) which are a natural outgrowth of the material covered in the previous chapters in this monograph. The topics range from data assimilation and real-time prediction for the MJO using the stochastic skeleton model, to the skeleton and "muscle" of the MJO and their interaction, to other MJO theories, and finally to stochastic parameterization to improve operational General Circulation Models (GCMs).

7.1 Future Directions for Stochastic Skeleton Models

7.1.1 Data Assimilation and Real-Time Prediction

Given the success of the stochastic model discussed in Chaps. 3, 5, and 6 in capturing the various types of MJO wave trains and detailed structure of the MJO in observations, it is a natural project to build a practical prediction scheme based on these models. There is a great advantage of such a method: a large number of prediction samples can be used with such a stochastic skeleton model which has only a few hundred degrees of freedom, thus overcoming the "curse of ensemble size" (Majda 2016). For real-time prediction, an accurate method for data assimilation, also called state estimation or filtering, is a crucial step. Such an algorithm has been developed recently in full detail for the stochastic skeleton model and validated on stringent tests (Chen and Majda 2016). The algorithm is based on contemporary mathematical theory involving judicious model error and conditional Gaussian structures in nonlinear systems (see Chapter 5 of Majda (2016) and Majda and Harlim (2012) for introductory background discussion).

© The Author(s), under exclusive licence to Springer Nature Switzerland AG 2019 113
A. J. Majda et al., *Tropical Intraseasonal Variability and the Stochastic Skeleton Method*, SpringerBriefs in Mathematics of Planet Earth,
https://doi.org/10.1007/978-3-030-22247-5_7

7.1.2   Rigorous Stochastic Attractors

As discussed in Chap. 3 there is a rigorous theory for the geometric ergodicity of the basic stochastic skeleton model (Majda and Tong 2015). Interesting research problems include generalization to the stochastic models with refined vertical structure from Chap. 6, and the development of a stochastic skeleton model with tropical–extratropical interaction based on Chaps. 3 and 4. These are both physically and mathematically interesting projects. A recent general framework for geometric ergodicity should be useful here (Tong and Majda 2016).

7.1.3   Skeleton Models and the Monsoon

For readers interested in stochastic skeleton models and the monsoon, please see Thual et al. (2015) and Ogrosky et al. (2017). Lack of space prevents detailed discussion.

7.2   The Skeleton and "Muscle" of the MJO

The salient features of the MJO in observations have been listed as the five items (I)–(V) in the introduction chapter and emphasized as a main goal of the skeleton models in the subsequent chapters of this monograph. While these are the salient planetary-intraseasonal features of MJO composites, individual MJO events often have additional features, such as westerly wind bursts (Lin and Johnson 1996; Majda and Biello 2004; Biello and Majda 2005; Majda and Stechmann 2009a), complex vertical structures (Lin and Johnson 1996; Myers and Waliser 2003; Kikuchi and Takayabu 2004; Kiladis et al. 2005; Tian et al. 2006), and complex convective features within the MJO envelope (Nakazawa 1988; Hendon and Liebmann 1994; Dunkerton and Crum 1995; Yanai et al. 2000; Houze et al. 2000; Masunaga et al. 2006; Kiladis et al. 2009). Since these additional features add detailed character to each MJO's structure, and since these features often account for additional strength beyond the MJO's skeleton, they are referred to here as the MJO's "muscle" (Majda and Stechmann 2009b). These effects of the muscle on the MJO have significant societal impact due to extreme weather associated with them (Lau and Waliser 2012). They lie in the envelope of convective activity including convectively coupled equatorial waves (CCEWs) and squall lines often involving complex multiscale interactions which often are elucidated by multi-scale mathematical models (Khouider et al. 2013). There are many challenging open problems here (Majda and Yang 2016; Yang and Majda 2017).

7.3   Other Models for the MJO

Many models for the MJO have been proposed, although none has been generally accepted. The present book herein describes a summary of the authors' body of work and views on the MJO. We next describe some other models for the MJO.

There have been a large number of theories attempting to explain the MJO through mechanisms such as evaporation–wind feedback (also known as wind-induced surface heat exchange, or WISHE) (Emanuel 1987; Neelin et al. 1987), boundary-layer frictional convergence instability (BFCI) (Wang and Rui 1990), stochastic linearized convection (Salby et al. 1994), radiation instability (Raymond 2001), and the planetary-scale linear response to moving heat sources (Chao 1987). In the present authors' opinion, these theories are all at odds with the observational record in various crucial ways (Zhang 2005; Lau and Waliser 2012), and it is therefore likely that none of them captures the fundamental physical mechanisms of the MJO. Nevertheless, they all provide some insight into different processes related to the MJO.

We briefly discuss two prominent examples of these theories, BFCI and WISHE. BFCI provides detailed prediction of the eastward slow MJO phase speed, but it predicts a Kelvin wave structure rather than the observed quadrupole vortex in (III). The phase speed in (I) is obtained by "tuning" the boundary layer thickness to unrealistically large values (Moskowitz and Bretherton 2000; Biello and Majda 2006). The WISHE theory is based on the fact that surface evaporation of water vapor has a nonlinear dependence on surface wind speed—an established observational fact at small scales. The original WISHE theory (Emanuel 1987) required easterly mean winds to tune the WISHE instability mode to have a moist Kelvin wave with an MJO phase speed in (I); this is contrary to the fact that most MJOs start in the Indian Ocean where the mean wind is not easterly but westerly. There is a more recent semi-empirical WISHE model (Sobel and Maloney 2012) which attempts to overcome the problem. No one doubts that the nonlinear dependencies of the surface evaporation can enhance convective activity; however, the central issue is whether this is a primary mechanism needed to generate a realistic MJO. In the present authors' opinion, the overwhelming evidence is no. First in Chaps. 2 and 3 above we have produced stochastic skeleton models with the detailed observational features (I)–(V) without any WISHE mechanisms. Furthermore, a series of recent careful numerical simulations produces very realistic MJO initiation and propagation through multicloud parameterization with the WISHE mechanism absent and the surface evaporation coefficient constant (Khouider et al. 2011; Ajayamohan et al. 2013, 2014; Deng et al. 2015, 2016).

7.4   Improving the MJO and Monsoon in GCMs Through Judicious Stochastic Parameterization

A central theme of this monograph from Chaps. 3 and 6 is that the use of judicious stochastic parameterization of the envelope of convective activity leads to more realistic behavior for MJOs beyond the deterministic model in Chap. 2. A similar but more complex strategy shows great promise over conventional deterministic cloud parameterization (CP) to greatly improve TISV for MJOs and monsoons in operational GCMs. The deterministic nature of a CP scheme smooths out the convective variability for subgrid dimensions (Arakawa 2004; Peters et al. 2013). This limitation of the deterministic CP schemes got further exposed when stochastic (Buizza et al. 1999; Lin and Neelin 2000, 2002, 2003; Palmer 2001; Majda and Khouider 2002; Khouider et al. 2003; Plant and Craig 2008; Teixeira and Reynolds 2008; Khouider et al. 2010) and cloud resolving (Grabowski and Smolarkiewicz 1999; Grabowski 2001; Khairoutdinov and Randall 2001; Randall et al. 2003; Satoh et al. 2008; Fudeyasu et al. 2008; Benedict and Randall 2009; Liu et al. 2009) approaches showed promising improvements. While superparameterized and global cloud-resolving models continue to evolve (Goswami et al. 2015; Yashiro et al. 2016; Fukutomi et al. 2016; Kooperman et al. 2016), these approaches are computationally very expensive. Stochastic approaches are getting more and more attention in recent times (Deng et al. 2015, 2016; Ajayamohan et al. 2016; Davini et al. 2017; Goswami et al. 2017a; Dorrestijn et al. 2016; Wang et al. 2016; Berner et al. 2017; Peters et al. 2017) as a computationally cheap alternative.

Recent studies involving the stochastic multicloud model (SMCM) (Khouider et al. 2010, 2011; Ajayamohan et al. 2013, 2014, 2016; Deng et al. 2015, 2016, for example) have shown considerable promise in simulating TISV. See chapters above for the basic ideas in SMCM. For a moist background with moderate stratiform fraction (or a drier background with large stratiform fraction), Khouider et al. (2011) demonstrated a course-resolution GCM could simulate many observed features of TISV, like the MJO and CCEWs. Ajayamohan et al. (2013) addressed MJO initiation and dynamics through realistic simulation of circumnavigating Kelvin waves. Deng et al. (2015, 2016) showed that SMCM incorporated into the National Center for Atmospheric Research's (NCAR) High-Order Methods Modeling Environment (HOMME) model in an aquaplanet setup produces realistic MJOs with front-to-rear vertical tilt and quadrupole vortex structure (Kiladis et al. 2005). Ajayamohan et al. (2014, 2016) showed that the simulation of the Indian summer monsoon ISOs can be improved by incorporating SMCM in the NCAR–HOMME model. However, all of the above results are based on idealized model simulations. Therefore, implementing SMCM into a fully coupled climate model was an obvious way forward. A group took up the NCEP CFSv2 model, promoted by the National Monsoon Mission of the Ministry of Earth Sciences, India, and implemented the SMCM into it. Some preliminary results of the implementation have appeared in Goswami et al. (2017a). The study by Goswami et al. (2017b) is an extension to Goswami et al. (2017a) in the sense that it provides in-depth analysis of intraseasonal variability of the CFS-SMCM-simulated climate whose mean state has briefly been described in Goswami et al. (2017a).

The SMCM is, at this stage, an experimental parameterization approach (Khouider and Majda 2006; Khouider et al. 2010) where the convective heating is parameterized based on three prescribed basis functions (see Chap. 6). These basis functions are designed to mimic the three major cloud types of tropical convection, namely congestus, deep, and stratiform (Johnson et al. 1999; Mapes et al. 2006). SMCM divides each CFSv2 grid box into a 40 × 40 microscopic lattice. Each lattice site is occupied by either congestus, deep, or stratiform cloud decks, or it is considered a clear sky site. Transitions from a lattice site with one type of cloud to another type occur according to a stochastic Markov chain process with transition probabilities that are conditioned to the large-scale convective available potential energy (CAPE), convective inhibition (CIN), middle tropospheric (700 hPa) dryness/moistness (MTD), and vertical velocity at the top of the boundary layer (W). Each microscopic lattice within a large-scale (read CFSv2) grid box sees the same large scale conditions. However, their evolutions in time differ as the transition rules depend on the previous state of a microscopic lattice as well as the large-scale conditions. Heating due to three cloud types is parameterized to be proportional to the area fractions occupied by the respective cloud types. The three prescribed basis functions of the SMCM are amplified by the respective parameterized heating, and the amplified profiles add up to yield the total parameterized heating. The moisture and temperature tendencies are calculated from this parameterized total heating and then given back to the host model, which is CFSv2.

The parameters in the SMCM have been chosen originally from physical intuition but were subsequently estimated from observations (Peters et al. 2013) and idealized simulations through systematic Bayesian statistical inference (De La Chevrotière et al. 2014, 2016) with very good agreement.

References

Ajayamohan RS, Khouider B, Majda AJ (2013) Realistic initiation and dynamics of the Madden–Julian oscillation in a coarse resolution aquaplanet GCM. Geophys Res Lett 40(23):6252–6257

Ajayamohan RS, Khouider B, Majda AJ (2014) Simulation of monsoon intraseasonal oscillations in a coarse-resolution aquaplanet GCM. Geophys Res Lett 41(15):5662–5669

Ajayamohan RS, Khouider B, Majda AJ, Deng Q (2016) Role of stratiform heating on the organization of convection over the monsoon trough. Clim Dyn 47(12):3641–3660

Arakawa A (2004) The cumulus parameterization problem: past, present, and future. J Clim 17(13):2493–2525

Benedict J, Randall D (2009) Structure of the Madden–Julian oscillation in the superparameterized CAM. J Atmos Sci 66(11):3277–3296

Berner J, Achatz U, Batte L, Bengtsson L, Cámara Adl, Christensen HM, Colangeli M, Coleman DRB, Crommelin D, Dolaptchiev SI et al (2017) Stochastic parameterization: toward a new view of weather and climate models. Bull Am Meteorol Soc 98(3):565–588

Biello JA, Majda AJ (2005) A new multiscale model for the Madden–Julian oscillation. J Atmos Sci 62:1694–1721

Biello JA, Majda AJ (2006) Modulating synoptic scale convective activity and boundary layer dissipation in the IPESD models of the Madden–Julian oscillation. Dyn Atmos Oceans 42:152–215

Buizza R, Miller M, Palmer TN (1999) Stochastic representation of model uncertainties in the ECMWF ensemble prediction system. Q J Roy Meteorol Soc 125(560):2887–2908

Chao WC (1987) On the origin of the tropical intraseasonal oscillation. J Atmos Sci 44:1940–1949

Chen N, Majda AJ (2016) Filtering the stochastic skeleton model for the Madden–Julian oscillation. Mon Weather Rev 144(2):501–527

Davini P, von Hardenberg J, Corti S, Christensen HM, Juricke S, Subramanian A, Watson PAG, Weisheimer A, Palmer TN (2017) Climate SPHINX: evaluating the impact of resolution and stochastic physics parameterisations in the EC-Earth global climate model. Geosci Model Dev 10(3):1383

De La Chevrotière M, Khouider B, Majda AJ (2014) Calibration of the stochastic multicloud model using bayesian inference. SIAM J Sci Comput 36(3):B538–B560

De La Chevrotière M, Khouider B, Majda AJ (2016) Stochasticity of convection in Giga-LES data. Clim Dyn 47(5–6):1845–1861

Deng Q, Khouider B, Majda AJ (2015) The MJO in a coarse-resolution GCM with a stochastic multicloud parameterization. J Atmos Sci 72:55–74. https://doi.org/10.1175/JAS-D-14-0120.1

Deng Q, Khouider B, Majda AJ, Ajayamohan RS (2016) Effect of stratiform heating on the planetary-scale organization of tropical convection. J Atmos Sci 73(1):371–392

Dorrestijn J, Crommelin DT, Siebesma AP, Jonker HJJ, Selten F (2016) Stochastic convection parameterization with Markov chains in an intermediate-complexity GCM. J Atmos Sci 73(3):1367–1382

Dunkerton TJ, Crum FX (1995) Eastward propagating $\sim$2- to 15-day equatorial convection and its relation to the tropical intraseasonal oscillation. J Geophys Res 100(D12):25781–25790

Emanuel KA (1987) An air–sea interaction model of intraseasonal oscillations in the Tropics. J Atmos Sci 44:2324–2340

Fudeyasu H, Wang Y, Satoh M, Nasuno T, Miura H, Yanase W (2008) Global cloud-system-resolving model NICAM successfully simulated the lifecycles of two real tropical cyclones. Geophys Res Lett 35(22):L22808

Fukutomi Y, Kodama C, Yamada Y, Noda AT, Satoh M (2016) Tropical synoptic-scale wave disturbances over the western Pacific simulated by a global cloud-system resolving model. Theor Appl Climatol 124(3–4):737–755

Goswami BB, Krishna RPM, Mukhopadhyay P, Khairoutdinov M, Goswami BN (2015) Simulation of the Indian summer monsoon in the superparameterized climate forecast system version 2: preliminary results. J Clim 28(22):8988–9012

Goswami BB, Khouider B, Phani R, Mukhopadhyay P, Majda A (2017a) Improving synoptic and intraseasonal variability in CFSv2 via stochastic representation of organized convection. Geophys Res Lett 44(2):1104–1113

Goswami BB, Khouider B, Phani R, Mukhopadhyay P, Majda AJ (2017b) Improved tropical modes of variability in the NCEP Climate Forecast System (Version 2) via a stochastic multicloud model. J Atmos Sci 74(10):3339–3366

Grabowski WW (2001) Coupling cloud processes with the large-scale dynamics using the cloud-resolving convection parameterization (CRCP). J Atmos Sci 58:978–997

Grabowski WW, Smolarkiewicz PK (1999) CRCP: a cloud resolving convection parameterization for modeling the tropical convecting atmosphere. Phys D Nonlinear Phenomena 133:171–178

Hendon HH, Liebmann B (1994) Organization of convection within the Madden–Julian oscillation. J Geophys Res 99:8073–8084. https://doi.org/10.1029/94JD00045

Houze RA Jr, Chen SS, Kingsmill DE, Serra Y, Yuter SE (2000) Convection over the Pacific warm pool in relation to the atmospheric Kelvin–Rossby wave. J Atmos Sci 57:3058–3089

Johnson RH, Rickenbach TM, Rutledge SA, Ciesielski PE, Schubert WH (1999) Trimodal characteristics of tropical convection. J Clim 12:2397–2418

Khairoutdinov MF, Randall DA (2001) A cloud resolving model as a cloud parameterization in the NCAR Community Climate System Model: preliminary results. Geophys Res Lett 28(18):3617–3620

Khouider B, Majda AJ (2006) A simple multicloud parameterization for convectively coupled tropical waves. Part I: linear analysis. J Atmos Sci 63:1308–1323

Khouider B, Majda AJ, Katsoulakis MA (2003) Coarse-grained stochastic models for tropical convection and climate. Proc Natl Acad Sci USA 100(21):11941–11946

Khouider B, Biello JA, Majda AJ (2010) A stochastic multicloud model for tropical convection. Commun Math Sci 8:187–216

Khouider B, St-Cyr A, Majda AJ, Tribbia J (2011) The MJO and convectively coupled waves in a coarse-resolution GCM with a simple multicloud parameterization. J Atmos Sci 68:240–264

Khouider B, Majda AJ, Stechmann SN (2013) Climate science in the tropics: waves, vortices and PDEs. Nonlinearity 26(1):R1–R68

Kikuchi K, Takayabu YN (2004) The development of organized convection associated with the MJO during TOGA COARE IOP: trimodal characteristics. Geophys Res Lett 31(10.1029):L10101

Kiladis GN, Straub KH, Haertel PT (2005) Zonal and vertical structure of the Madden–Julian oscillation. J Atmos Sci 62:2790–2809

Kiladis GN, Wheeler MC, Haertel PT, Straub KH, Roundy PE (2009) Convectively coupled equatorial waves. Rev Geophys 47:RG2003. https://doi.org/10.1029/2008RG000266

Kooperman GJ, Pritchard MS, Burt MA, Branson MD, Randall DA (2016) Robust effects of cloud superparameterization on simulated daily rainfall intensity statistics across multiple versions of the Community Earth System Model. J Adv Model Earth Syst 8(1):140–165

Lau WKM, Waliser DE (eds) (2012) Intraseasonal Variability in the atmosphere–ocean climate system, 2nd edn. Springer, Berlin

Lin X, Johnson RH (1996) Kinematic and thermodynamic characteristics of the flow over the western Pacific warm pool during TOGA COARE. J Atmos Sci 53:695–715

Lin J, Neelin J (2000) Influence of a stochastic moist convective parameterization on tropical climate variability. Geophys Res Lett 27(22):3691–3694. https://doi.org/10.1029/2000GL011964

Lin J, Neelin J (2002) Considerations for stochastic convective parameterization. J Atmos Sci 59(5):959–975

Lin J, Neelin J (2003) Toward stochastic deep convective parameterization in general circulation models. Geophys Res Lett 30(4):1162. https://doi.org/10.1029/2002GL016203

Liu P, Satoh M, Wang B, Fudeyasu H, Nasuno T, Li T, Miura H, Taniguchi H, Masunaga H, Fu X, et al (2009) An MJO simulated by the NICAM at 14-and 7-km resolutions. Mon Weather Rev 137(10):3254–3268

Majda AJ (2016) Introduction to turbulent dynamical systems in complex systems. Springer, Berlin

Majda AJ, Biello JA (2004) A multiscale model for the intraseasonal oscillation. Proc Natl Acad Sci USA 101(14):4736–4741

Majda AJ, Harlim J (2012) Filtering turbulent complex systems. Cambridge University Press, Cambridge

Majda A, Khouider B (2002) Stochastic and mesoscopic models for tropical convection. Proc Natl Acad Sci USA 99(3):1123–1128

Majda AJ, Stechmann SN (2009a) A simple dynamical model with features of convective momentum transport. J Atmos Sci 66:373–392

Majda AJ, Stechmann SN (2009b) The skeleton of tropical intraseasonal oscillations. Proc Natl Acad Sci USA 106(21):8417–8422

Majda AJ, Tong XT (2015) Geometric ergodicity for piecewise contracting processes with applications for tropical stochastic lattice models. Commun Pure Appl Math. https://doi.org/10.1002/cpa.21584

Majda AJ, Yang Q (2016) A multiscale model for the intraseasonal impact of the diurnal cycle over the maritime continent on the Madden–Julian oscillation. J Atmos Sci 73(2):579–604

Mapes BE, Tulich S, Lin JL, Zuidema P (2006) The mesoscale convection life cycle: building block or prototype for large-scale tropical waves? Dyn Atmos Oceans 42:3–29

Masunaga H, L'Ecuyer T, Kummerow C (2006) The Madden–Julian oscillation recorded in early observations from the Tropical Rainfall Measuring Mission (TRMM). J Atmos Sci 63(11):2777–2794

Moskowitz BM, Bretherton CS (2000) An analysis of frictional feedback on a moist equatorial Kelvin mode. J Atmos Sci 57(13):2188–2206

Myers D, Waliser D (2003) Three-dimensional water vapor and cloud variations associated with the Madden–Julian oscillation during Northern Hemisphere winter. J Clim 16(6):929–950

Nakazawa T (1988) Tropical super clusters within intraseasonal variations over the western Pacific. J Meteorol Soc Jpn 6(6):823–839

Neelin JD, Held IM, Cook KH (1987) Evaporation–wind feedback and low-frequency variability in the tropical atmosphere. J Atmos Sci 44:2341–2348

Ogrosky HR, Stechmann SN, Majda AJ (2017) Boreal summer intraseasonal oscillations in the MJO skeleton model with observation-based forcing. Dyn Atmos Oceans 78:38–56

Palmer TN (2001) A nonlinear dynamical perspective on model error: a proposal for non-local stochastic-dynamic parametrization in weather and climate prediction models. Q J Roy Meteorol Soc 127(572):279–304

Peters K, Jakob C, Davies L, Khouider B, Majda AJ (2013) Stochastic behavior of tropical convection in observations and a multicloud model. J Atmos Sci 70(11):3556–3575

Peters K, Crueger T, Jakob C, Möbis B (2017) Improved MJO simulation in ECHAM6.3 by coupling a stochastic multicloud model to the convection scheme. J Adv Model Earth Syst 9(1):193–219

Plant R, Craig G (2008) A stochastic parameterization for deep convection based on equilibrium statistics. J Atmos Sci 65(1):87–105

Randall D, Khairoutdinov M, Arakawa A, Grabowski W (2003) Breaking the cloud parameterization deadlock. Bull Am Meteorol Soc 84:1547–1564

Raymond DJ (2001) A new model of the Madden–Julian oscillation. J Atmos Sci 58:2807–2819

Salby ML, Garcia RR, Hendon HH (1994) Planetary-scale circulations in the presence of climatological and wave-induced heating. J Atmos Sci 51:2344–2367

Satoh M, Matsuno T, Tomita H, Miura H, Nasuno T, Iga SI (2008) Nonhydrostatic icosahe-dral atmospheric model (NICAM) for global cloud resolving simulations. J Comput Phys 227(7):3486–3514

Sobel A, Maloney E (2012) An idealized semi-empirical framework for modeling the Madden–Julian oscillation. J Atmos Sci 69(5):1691–1705

Teixeira J, Reynolds CA (2008) Stochastic nature of physical parameterizations in ensemble prediction: a stochastic convection approach. Mon Weather Rev 136(2):483–496

Thual S, Majda AJ, Stechmann SN (2015) Asymmetric intraseasonal events in the stochastic skeleton MJO model with seasonal cycle. Clim Dyn 45:603–618

Tian B, Waliser D, Fetzer E, Lambrigtsen B, Yung Y, Wang B (2006) Vertical moist thermodynamic structure and spatial–temporal evolution of the MJO in AIRS observations. J Atmos Sci 63(10):2462–2485

Tong XT, Majda AJ (2016) Moment bounds and geometric ergodicity of diffusions with random switching and unbounded transition rates. Res Math Sci 3(1):41

Wang B, Rui H (1990) Dynamics of the coupled moist Kelvin–Rossby wave on an equatorial beta-plane. J Atmos Sci 47:397–413

Wang Y, Zhang GJ, Craig GC (2016) Stochastic convective parameterization improving the simulation of tropical precipitation variability in the NCAR CAM5. Geophys Res Lett 43(12):6612–6619

Yanai M, Chen B, Tung WW (2000) The Madden–Julian oscillation observed during the TOGA COARE IOP: global view. J Atmos Sci 57:2374–2396

Yang Q, Majda AJ (2017) Upscale impact of mesoscale disturbances of tropical convection on synoptic-scale equatorial waves in two-dimensional flows. J Atmos Sci 74(9):3099–3120

Yashiro H, Kajikawa Y, Miyamoto Y, Yamaura T, Yoshida R, Tomita H (2016) Resolution dependence of the diurnal cycle of precipitation simulated by a global cloud-system resolving model. SOLA 12:272–276

Zhang C (2005) Madden–Julian Oscillation. Rev Geophys 43:RG2003. https://doi.org/10.1029/2004RG000158

Index

A

Amplitude of the convection/wave activity envelope, 10

B

Background vertical moisture gradient, 11, 44, 52, 75
Baroclinic modes
  first, 11, 52, 69, 104
  second, 101, 107–108, 110
Baroclinic potential temperature, 72, 103, 107
Baroclinic velocity, 52, 103, 107
Barotropic pressure, 51–52
Barotropic Rossby waves, 49, 51, 56–60, 64
Barotropic velocity, 51–52
Beta-plane approximation, 18
Beta-plane model, 18
Boundary-layer frictional convergence instability (BFCI), 115
Boussinesq equations, 52

C

CCEWs, *see* Convectively couple equatorial waves
Characteristic variables, 15–17, 87, 90
Circumnavigating MJO, 35, 36, 116
Convective heating and drying, 50, 67, 80, 82, 117
Convectively couple equatorial waves (CCEWs), 21, 30, 55, 56, 68, 114, 116
Coriolis force, 52

D

Data assimilation, 29, 113
2-Day waves, 30
Dispersion relation
  of convectively coupled equatorial waves, 55
  of the MJO, 6, 18, 19, 46
  of the model (amplitude-dependent), 23

E

El Nino southern oscillation (ENSO), 1, 5, 49, 67, 69, 74
Empirical orthogonal functions (EOFs), 68, 84, 90
Energy
  conserved total energy, 52
  conserved vertically integrated moist static energy of the skeleton model, 13, 106, 108, 109
  convective energy, 13, 52
  dry baroclinic energy, 52
  dry barotropic energy, 52
  moisture energy, 52
ENSO, *see* El Nino southern oscillation
EOFs, *see* Empirical orthogonal functions
Equatorial long-wave scaling/approximation, 11, 16, 68, 85–90, 102–103
Equatorial Rossby waves
  dry, 12
  long-wave, 12, 70
  moist, 21
Equatorial wave variables, 14, 16

© The Author(s), under exclusive licence to Springer Nature Switzerland AG 2019
A. J. Majda et al., *Tropical Intraseasonal Variability and the Stochastic Skeleton Method*, SpringerBriefs in Mathematics of Planet Earth,
https://doi.org/10.1007/978-3-030-22247-5

F
Fourier series, 45
Froude number, 55

G
Galerkin truncation, 52
General circulation model (GCM), 1, 6, 36, 94,
     95, 113, 116–117
Geopotential height, 72, 73
Geostrophic balance (meridional geostrophic
     balance), 16, 17, 85–90
Gillespie algorithm, 41–42, 44

H
Hamiltonian, 42–44
Heating/drying rate, 11, 44, 75, 80
Hermite polynomials, 14, 71
Hessian matrix, 56
Horizontal quadrupole vortices, 2, 6, 19
Hydrostatic balance equation, 73

I
Interpolated outgoing OLR, 72
Intraseasonal oscillation frequency, 18, 49, 64
Intraseasonal planetary band/MJO band, 6, 7,
     18, 31, 33, 99

K
Kelvin waves
     dry, 21, 33, 36, 55, 78, 81, 109
     moist, 115
Korteweg-de Vries equation, 24

L
La Nina, 74
Long-wave equatorial primitive equations, 70
Lower tropospheric moisture, 7, 9, 10, 19, 20,
     67, 75, 77

M
Madden-Julian oscillation (MJO)
     initiation and termination, 57, 58, 115, 116
     'muscle' features, 95
     'skeletal' features, 5–24, 32–39, 95
Markov birth-death process, 39–44
Master equation, 40, 41
Matsuno-Gill model, 12, 69, 104
Mean-field equation, 40, 41

Meridional geostrophic imbalance (MGI), 88,
     89
Mesoscale processes, 29, 30, 95
Mixed Rossby-gravity (MRG) waves, 16
MJO skeleton model
     deterministic form, 5–24, 34, 36, 42, 45
     extratropical-tropical form, 49–64
     refined vertical structure form, 93–110, 114
     stochastic full form, 29–46
     stochastic single-column form, 42–43, 45
MJO skeleton signal (MJOS), 75–79, 83, 84
MJO skeleton signal amplitude (MJOSA), 58,
     77, 78, 82, 85
MJO skeleton signal phase (MJOSP), 84
MJO wave trains, 2, 29, 31–36, 39, 42, 43, 46,
     62, 99, 101, 109, 113
Multicloud model, 95, 96, 98, 116

N
Noise-induced chaos, 42, 44
Nonlinear oscillator model, 7, 9, 75

O
OLR MJO Index (OMI), 84
Ordinary differential equations (ODEs), 23,
     24, 51, 56–58, 60, 63
Outgoing longwave radiation (OLR), 37, 50,
     70, 72, 76, 79, 80, 84, 90, 94

P
Parabolic cylinder functions, 13, 14, 16, 17,
     45, 70
Partial differential equations (PDEs), 2, 56, 67,
     75, 90
Planetary-scale circulation anomalies, 1, 5
Poisson law, 42
Potential temperature, 11, 32, 33, 36–38, 44,
     52, 69, 73, 98, 103, 107
Power spectra, 36–37, 39, 99–101, 103, 109
Primitive equations, 11, 50, 52, 70
Primitive variables, 14, 80

R
Radiative-convective equilibrium (RCE), 44,
     52, 54, 97, 103
Radiative cooling, 52, 54, 59
Raising and lowering operators, 15
RCE, *see* Radiative-convective equilibrium
Real-time multivariate MJO (RMM) index, 79,
     84
Reanalysis data, 73, 75–77, 79, 83, 90

Relative meridional geostrophic imbalance
    (RMGI), 89
Resonance conditions, 57–60, 63, 64
Resonant wave interactions, 49, 51, 56, 58–59,
    64
Rigid-lid boundary, 52
RMM index, *see* Real-time multivariate MJO
    index

S
Seasonal cycle, 38
Sea surface temperature (SST), 13, 21, 39, 74
Shallow-water equations, 69
Single-column skeleton model, 42–44
SMCM, *see* Stochastic multicloud model
Soliton solutions, 24
Spatiotemporal filtering, 20, 89
SST, *see* Sea surface temperature
Standing oscillations, 21
Stochastic multicloud model (SMCM), 116,
    117

Streamfunction, 53
Synoptic scale processes, 10, 19, 29, 30, 32,
    36, 39, 56, 79, 95, 96, 110

T
Three-wave triads, 57
Topographic resonance, 60
Transition rates, 40–42, 98, 106
Traveling waves, 23–24
Tropical-extratropical interactions, 49–64, 114
Two-layer equatorial beta-plane equations,
    51–53

W
Walker circulation, 49, 59–62, 64, 67–90
Warm pool, 1, 13, 21, 34, 35, 37–39, 45, 79–85
Westerly wind bursts, 6, 19, 94, 114
Wind-induced surface heat exchange (WISHE),
    115
Wind shear, 62–64